Cd

LONDON'S ENVIRONMENT

Prospects for a Sustainable World City

LONDON'S

ENVIRONMENT

Prospects for a Sustainable World City

edited by Julian Hunt

University College London, UK

Imperial College Press

Published by

Imperial College Press
57 Shelton Street
Covent Garden
London WC2H 9HE

Distributed by

World Scientific Publishing Co. Pte. Ltd.

5 Toh Tuck Link, Singapore 596224

USA office: 27 Warren Street, Suite 401-402, Hackensack, NJ 07601

UK office: 57 Shelton Street, Covent Garden, London WC2H 9HE

British Library Cataloguing-in-Publication Data
A catalogue record for this book is available from the British Library.

LONDON'S ENVIRONMENT
Prospects for a Sustainable World City

ISBN   1-86094-486-8

Editor: Tjan Kwang Wei

Cover Photos: Dr Andrew Hudson-Smith, Centre for Advanced Spatial Analysis,
               University College London

Printed in Singapore by B & JO Enterprise

Preface

"For there is in London all that life can afford"

Samuel Johnson, 1777

An exciting and wide ranging conference on 'London's environment and future' was held at the Senate House in Bloomsbury, on September 18–19, 2002 with the academic support of University of London, University College London, Imperial College, London Metropolitan University and Middlesex University. Excellent lectures and other presentations were given by London's leading experts in architecture, planning, politics and environment including its impact on health and the arts. They included the then Minister of the Environment Michael Meacher, the Mayor of London Ken Livingstone and the architect and government advisor Lord Rogers. The 250 attending the conference came from every walk of life and every type of organisation including children from two of London's more environmentally conscious schools. Everyone participated actively in the two panel discussions on environment and sustainability, and in the breakout groups whose summarised views and specific conclusions were discussed in the closing session of the conference. A programme of songs and verse on Elemental London by the Weatherwise Theatre Company was presented in the Wellcome Wing of the Science Museum.

The authors of the nineteen chapters of this book, based on the conference lectures, hope that they are useful and enjoyable for reading by any general reader interested in the environment and in the long term sustainability of world cities generally, and London in particular. A postscript by the Deputy Mayor of London, Nicky Gavron, is included summarising London's environmental policies that have been developed since the conference was held. The articles have been edited to minimise jargon. Where necessary it is explained in the text and cross referenced in the index.

The views of the breakout groups and the leaders of the panel discussions are in the appendix, where there are also biographical details of all the authors.

The main organisers of the conference were Julian Hunt, Andrew Orr, Leisa Clemente, Roger Wotton, Carolyn Harrison, Phil Steadman, Roger Mackett, John Murlis, Heather Binney and Alan Lord of UCL, Helen ApSimon, Linda Davies, Roy Colvile, Tariq Ali and Sam Hunt of Imperial College, Peter Brittain of the Government Office for London, Chris Burnham of the Environment Agency, Jemima Hunt (journalist), Tony Fletcher, Carolyn Stephens and Mark McCarthy of the London School of Hygiene and Tropical Medicine, David Goode of GLA, Paul Henderson of the Natural History Museum, Alan Morton of the Science Museum and Phil Thompson of the Corporation of London. We are very grateful to the sponsors: DEFRA/Government Office for London, Environment Agency, Arup Environmental, Thames Water, Bloomsbury Institute for the Natural Environment, CPOM at University College London, Young's Brewery and Cobra Beer.

The idea of assembling this book came from colleagues at University College, Mike Batty and David Banister, who suggested that it might be useful for students in architecture, planning and environmental disciplines.

I am also grateful for the support and suggestions of my family, friends and colleagues in editing the book, especially Marylla, Jemima, Tilly and Tristram Hunt, Giles Foden, Sam Hunt, Phil Steadman, Andrew Orr and Leisa Clemente.

Contents

World Contexts

Chapter 1

London's Sustainability — An Overview

Julian Hunt

THE IDEA

Cities are amongst the greatest of human creations, though at the cost of consuming ever more of the Earth's energy, materials, and water and discharging them into the environment. Their governments, including London's, are very concerned that the future well-being of their cities is threatened both by these resources becoming limited and less plentiful in future, and by the increasing damage that large conurbations are causing to their environmental surroundings. In a major shift in strategic thinking city leaders have come to the conclusion that their future development plans will have to be based on the principle of sustainability; a concept inspired by the analogy between the ideal functioning of cities and successful ecological systems. Some pioneering smaller cities in both developing and developed countries, from Curitiba in Brazil to Woking in the UK, have demonstrated that the 'sustainability' approach has many practical benefits; populations should be more healthy and secure against disasters; the cities should have thriving and stable economies, balanced societies and, liveable conditions; for the future it is especially important that the resource requirements and impacts of the city should be in balance with the natural environment within and outside its boundaries. The test of 'sustainable development', as we now understand this term following Brundtland (1987), is essentially social and political, in that it should successfully continue for generations. In some ways this is an

3

old idea; the duty of benefiting future generations was expected of each citizen by the city states of ancient Greece.

Even fifty years ago less than one third of the global population lived in cities, though now urban areas contain more than half the world's population, which will rise to two-thirds by the middle of this century (UN 2001). A century ago in the UK most people lived in towns and cities and now this proportion is over 80% (DETR 2000). Inevitably, as urban areas use up ever more of the world's resources they also contribute most of the discharges to the environment. This is why conurbations directly and indirectly are largely responsible for most of the changes in the world's environment caused by human activities. Their global impacts and 'ecological' footprints are greater than those of many medium-sized countries. Perhaps, with some justice, urban communities have become especially vulnerable to the effects of these changes, whether originating in the atmosphere, or the ocean or the land areas around them. Most of the largest urban areas with populations of many millions, whether in Europe, America, Asia or Africa, are on low-lying coasts and estuaries. Here they are particularly subject to the dangers of floods and rising sea levels, associated with the expected changes in the world's climate. (Chapter 13, Hunt 2005)

The tasks of governing and managing these world cities and the risks they face in rapidly changing economic, social, and environmental conditions are hugely complex and demanding. Overarching principles are needed to guide political leadership and planning. They should be sufficiently clear and obviously beneficial that, with advocacy, the population can understand and participate in these developments. There is now firm survey evidence which validates this political aspect of sustainability; an outcome which has surprised even those who first advocated this approach (Jones 2004). For these reasons many cities have adopted the goal of moving towards sustainability as the best guide for ensuring that management and planning decisions are both consistent and 'joined up', especially those affecting environmental issues. Academic studies in urban development have contributed critiques that, as one would expect, have shown inconsistencies and pitfalls; but more positively they have also shown how the sustainability concept can be extended, for example down to smaller community or commercial scale developments and up to the scale of whole countries (Chapter 16). The first relevant Act of the UK Parliament (HMG 1995) was the Environment Act in 1995, which required the Environment Agency to work towards the objectives of sustainable development. A recent Act in 2004, showing how this concept is now firmly embedded in legislation, requires adherence to this principle in decisions on planning and economic regeneration. (In the passage of this and related Bills there have been continuing debates about what sustainability means, but in practice the principle itself is now generally accepted.) In London, the Mayor has established, following widespread consultation, detailed objectives and programmes of action over many areas of policy. These are being coordinated within an overall framework of sustainable development. He has emphasised that the global responsibility of a world city means that it should be exemplary in taking its share of the actions needed to ensure the future sustainability of our planet.

The authors of this book have contributed authoritative and original insights into the political, social, economic, administrative and scientific aspects of the changing environment of London. They have particularly focussed on the changes caused by environmental pressures associated with developments in building, transport, technology, and people's behaviour and expectations. All these discussions also have to take into account the future longer term variations in the global climate that will seriously affect London within the next 30 to 50 years (Chapters 3, 5, 19). A common theme in most of these chapters is that from both practical and academic points of view sustainability is the relevant concept for the broad analysis and assessment of the different aspects of London's present and future environment. It is not only a basis for environmental legislation but also, despite ambiguities in its interpretation, a framework for deciding on practical actions. Intrinsic to this approach is a careful consideration of how the environment is likely to evolve over time. In several chapters references to the history of London's environment provide valuable perspectives of past environmental change. These enable us to reflect whether the controversial strategies now being considered, especially those covering London's growth, will safeguard its natural habitats and will secure the city against natural or other possible disasters.

THE NATURAL SETTING

London's geography and natural environment have greatly influenced its development throughout its history, and it continues to do so even when the city's own environment dominates over that of its natural setting. As we have already noted, the city may become more vulnerable to environmental dangers. Luke Howard (1833) was the first to show how measurements and scientific study enable the environment to be understood. Focussing on

London's atmosphere and pollution, he pointed out how it differs from its state in the surrounding areas. In emphasising the need for adequate monitoring of the weather by official bodies, he compared London unfavourably with Paris. Such data is even more necessary today and for many other environmental as well as social indicators. In our more quantitative age this kind of information is what citizens and governments rely on to know what is happening and for deciding what needs to be done for the future of the environment.

London lies between hills of about 100 to 200 m high, in a shallow basin that extends about 30 km East to West and 30 km North to South. The present shape of the built up area has become more circular as it has spread over the hills. The valley was formed by the meandering river Thames flowing east to the North Sea through the shoals, sandbanks and coastal marshes of the widening estuary, where the river becomes tidal (Chapter 12). Although London was settled by the Romans in a small area on the north bank of the Thames (Chapter 4), until the Middle Ages the city was surrounded by fields and woodland which provided much of its food and fuel. The river has a vital influence on London; firstly it can be a danger (Chapter 13). In the 13th century even Westminster Hall in the Houses of Parliament was sometimes flooded. Secondly the Thames was also a great source of food; there were so many fish in the river water that they had to be swept out of the Hall after such floods subsided (Aykroyd 2001). Of the North European rivers entering the North Sea, the Thames estuary was one of the richest. It is reported in Melville's great novel, Moby Dick, that the early Kings of England were quite as proprietorial about the riches in their coastal seas as they were of their forests; ambergris extracted from a sperm whale is still one symbolic element in the oil used in the monarch's coronation. Thirdly because of the city's inadequate roads and the swampy banks along the river, the Thames and its important tributaries provided the main means of transport. Like Venice and Stockholm, London's main centres of commerce and government are situated at the water side. These attracted first the Roman galleys and then Viking longboats to conquer and trade (Chapter 12). The Vikings also encamped upstream of the city at Fulham (Clout 2004).

In the 17th century the Thames enabled parliamentarians to have a ready escape route from an irate monarch. In the 19th century, because of their proximity to the river they had to breathe the same 'great stink' from untreated sewage that Londoners endured. Disraeli, the future Prime Minister, described the river as 'a stagnant pool reeking with ineffable and unbearable horror!' (T. Hunt 2004). As a result the great Victorian sewage system was constructed, that is still in use (see Chapter 18). In the 21st century Parliament shares with Londoners the risks of flooding, though they will be greatest in the Thames Gateway developments along the shores of the Thames Estuary, 15 km downstream, where roads, warehouses and houses will be built on a massive scale to accommodate London's growing population (Chapters 2, 13). These and other low lying areas are becoming more susceptible to the environmental changes that are occurring all over the world (Chapter 12).

This connection between the local environment and that of the whole world is now well understood by politicians (Chapters 3, 19). It is particularly the large urban areas in America, Asia and Europe, including London, that are contributing most to the rapid increase of greenhouse gases, with the consequential effects of global warming, increased river flows from episodes of higher rainfall, and thermal expansion of the oceans. In London and the east of the UK the effective sea level is further increased because the land is sinking at a comparable rate, leading to the total rise being as much as 0.8 m over the next 100 years. Parliamentarians may well be protected if the new flood defences for London now being considered are constructed in time. But it is very unlikely that a similar effectiveness of sea defences can be provided for all communities along the UK coastline (Hansard 2000).

The hills to the North and South of London tend to channel the prevailing westerly winds over the city. These used to drive the capital's windmills in the Middle Ages (Ackroyd 2001). Was the recent demonstration wind turbine on the South Bank a harbinger of more wind energy for London in the future? The less frequent easterly winds, which are also funnelled into the London area by the coasts and the estuary, used to be carefully monitored by the wind gauge in the old Admiralty boardroom for signs that the Dutch fleet might invade up the Thames in the 17th century. We shall be more worried in future by the possible implication of such winds for flooding. By contrast, in stable atmospheric conditions and as the wind drops, the hills around the London basin can trap polluted air and fog particles very effectively. In the late 19th century Monet captured the extraordinary luminosity of these atmospheric events in his beautiful paintings of views over the Thames at Westminster (Chapter 8). But their consequences for the health of Londoners over that period were devastating. It was only the great smog of 1952, which killed 3500 people in a few days, that eventually galvanised Parliament to pass the Clean Air Act in 1956 and eliminated the worst effects of coal burning in UK cities. Today one of the main aims of air pollution policy is to reduce the invisible hazards of excess levels of nitrogen dioxide and ozone resulting mainly from motor vehicle emissions, although the products of combustion from domestic and industrial furnaces also contribute. It is the fine particles now that cause the most serious impact of air pollution on the health of Londoners. They are less visible than the large, sooty smog particles of the past, although they may still be seen as milky or hazy skies on the most polluted days when long-range transport of industrial pollutants from overseas combines with London's local emissions. As with previous environmental improvements this policy objective should succeed by a judicious combination of technology and regulations (Chapters 5, 7). In 2003, the Mayor of London introduced a 'congestion charge' for most private vehicles in the city centre. As Singapore had shown earlier, this successfully reduced traffic in the centre. In London it did not increase traffic or air pollution in the outer areas, both of which were considered to be possibilities based on some experiences in continental cities.

Although many aged and sick Londoners still succumb to cold winter temperatures in homes that are often damp and badly insulated, a serious environmental health danger in the future is likely to be caused by the effects of global warming during summer days (Department of Health 2000) and the 'heat-island' effect that further raises urban temperatures (Chapter 7). The frequency will increase when there are very warm temperatures above 30°C. Often these events will be associated with high levels of air pollution exacerbated by stable atmospheric conditions (UKCIP 2000). Mitigating such dangerous conditions, whose deadly affects have been experienced in Chicago in 1995 (Klinenberg 2002), Greece in 2001 and in France in 2003, will require appropriate policies of urban planning, housing and transport. Also public health measures will be necessary to combat the likely spread of tropical diseases in these conditions. Short term measures taken in response to warnings from long range weather forecasts of extreme conditions should in future enable communities to reduce the worst effects on vulnerable social groups (Hunt 2004). The founders of ancient Babylon and modern New Delhi showed how gardens and parks can moderate high temperatures in desert conditions. Surveys have shown that even in London there can be a significant drop in temperatures of 1–2°C in London's parks. This is a major reason, discussed in Chapter 11, for ensuring that both the expansion and the ecological management of urban green spaces, are central to the sustainable redevelopment of London's derelict 'brown field' sites (Chapter 2). As peak temperatures rise air conditioning may become the norm for houses, as it already is for many public buildings. Will the present social inequality of 'fuel poverty' be transformed into 'temperature poverty' later in this century, or will this be averted by provision of new kinds of ventilation and cooling systems in all types of housing? An environmentally damaging and expensive solution would be to incorporate air conditioning into houses as well as buildings which is already necessary in the tropical conditions of Hong Kong. This is just one indicator of how the lives of Londoners and the nature of the environmental issues might be radically different in 2100.

The geological processes of the London area need to be understood not only for how they formed its hills and rivers, but also for how they determined the sloping strata below the surface. These are nicely visible from the air in the quarries across the London Basin. The gravel on the hills lying above the clay layer below it provided water and rough grazing for early communities and formed the ponds on hillside parks and heath land, as John Constable depicted in his landscapes of London under grey cumulous clouds. Nowadays visitors and walkers can enjoy these open spaces where a wide variety of human and natural life is always on display, though the ecological health of some of the ponds is now threatened by their very popularity (Chapter 9).

The water for modern London comes from aquifers in the chalk formations in the basin and from the rivers feeding into the Thames. Continuing to supply and purify the huge volume required (which is about equal to the average rate of rain falling on London) will probably be managed without great difficulty (Chapter 10), even though the population is rising and the domestic use of water in households is beginning to approach the lavish levels of North America, or ancient Rome. At the same time the industrial uses of water (e.g. for brewing) is actually reducing especially from bore holes. This is contributing to the water table rising in central London, which has the serious consequence of saturating the foundations of buildings in low lying areas, eventually causing costly structural damage.

Lying above the chalk and below the gravel is the clay of the Thames Basin that provided a fertile soil for the agriculture and market gardens that fed London, and even in the Middle Ages exported food to other cities such as Oxford. A special feature of 'London Clay' is its characteristic mixture of silicaceous minerals and combustible oils so that it can be easily fired with the minimum of fuel. On the one hand this clay, especially in Bedfordshire to the North of London provided most of the bricks of London's buildings, and on the other provided a nearly impervious base for the large waste tips, where much of London's waste materials were consigned. These two uses are no longer in balance because the rate of clay and chalk removal for building is much less than the increasing volume of solid waste generated in the city. New land fill sites have had to be found outside the city typically on ground with little agricultural value, a policy that will not be allowed under EU regulations in future.

Indeed, dealing with waste is now recognised as probably the most immediate test of London's new commitment to sustainability. Political leaders are responding by enabling people to recycle more of their solid waste, and encouraging them to adopt this is as a perfectly practicable community goal, as continental cities have demonstrated for at least a decade. The Greater London Authority (GLA) is also stimulating new approaches for the financial, engineering and retail sectors to find ways to reduce, process and hopefully utilise waste to further reduce the environmental impact (Chapters 5, 19). New initiatives should emerge; in Belgium there are now plans to build ecological houses using materials extracted from domestic and industrial waste. Other solutions involve reprocessing waste so as to replenish land for agriculture and gardens.

THE DIVERSITY OF THE CITY'S ENVIRONMENT

World cities, with their openness to expanding international trade and movement of people, are bound to be affected by the increasing diversity of people in their communities. They also import a greater diversity of natural species, including the vectors that carry diseases. Large cities are catalysts for the spread of this diversity not just in their own countries but on a continental scale. They have a responsible role in their future development to deal with these challenges in a way that is consistent with their sustainability objectives. They should then be able to share their experiences with other cities and smaller communities. Greater London has a population of over seven million made up of many more community groups than any other city in the world, except perhaps New York. Children in London schools speak more than 200 languages (Chapter 16). All these communities are affected by and contribute to the development of the city in different ways. There are more species (about 300) of flowering plants within a few kilometres of St. Paul's Cathedral than in most parts of the British countryside and even more species (about 1500) over the whole urban area. Indeed London has some of its own species, notably the London Plane tree, a hybrid that has survived London's environment since the 17th century (Aykroyd 2001). Bird life is similarly diverse with up to 50 species recorded in a small area in central London.

Nevertheless, there is concern about the disappearance of some formerly widespread species from many areas; the rapidly declining number of sparrows being a particularly sore point for Londoners (see Chapter 9 and Appendix). On the other hand some forms of British or Asian wildlife are appearing in London that have not been seen for centuries, such as the fox, or even before, such as perokeets and crabs (Chapter 9).

Achieving the twin goals of diversity and sustainability in cities requires understanding the many complex ways in which the urban environment affects plant, animal and human life. Through the systematic and holistic thinking of biology, one can consider the structure and the functioning of living beings over the entire spectrum of sizes and mixtures of species, ranging from microbes up to the diverse communities of urban areas. In the same way policies for the future conservation and/or control of plants and animals in an urban area also cannot be decided piecemeal. Holistic policies should be based on assessments of the current state and understanding of the distributions, habitats, and varieties of all the significant species, i.e. their biodiversity (Chapters 5 and 9). With the stimulus of the Intergovernmental Convention on Biological Diversity agreed at the United Nations Environment Conference at Rio in 1992, managing biodiversity now provides the framework of the official policies on nature conservation. It has transformed the way that professional and amateur observers, the general public, and the many local and national societies of naturalists are now involved. This approach is encouraging a deeper appreciation and support for the natural environment; not just in the beautiful parks and other open spaces with which London is so well endowed, but also in other biologically rich areas such gardens, railway lines, canals, quarries, etc. This might be an example to other world cities many of which have tragically lost their open spaces in order to accommodate their burgeoning population. But they still have these informal open spaces teeming with biological diversity.

The variability of the health of the human communities across London is just as complex as that of plants and animals. It is of critical significance for deciding on social policies that we know how the general welfare, life expectancy and susceptibility to disease vary according to the location, economic state, ethnic origins, and gender of individuals and communities. Booth's (1903) comprehensive report of health and poverty first showed the needs for these studies. Modern research reported here (Chapter 6), shows that even with a National Health Service that was absent in Booth's day, it is still vitally necessary to study these factors individually and through all the ways in which they interact. There may well be other factors yet to be understood which also affect physical and mental health, such as the levels of air pollution and noise (Chapters 5, 14). The effects of future rises in peak temperatures have already been mentioned. As Chapter 6 shows the inequality in community health is still substantial. Some groups in London experience much worse environmental conditions than others and relatively higher levels of mortality and disease. Purely medical remedies are not enough. A sustainable health policy requires greater understanding of these connected issues by adversely affected communities and then their participation in local environmental and health programmes. As an experienced environmental official recently commented, these connections are not sufficiently appreciated by ethnic minorities in London, who are not much involved in environmental community groups and public bodies. By contrast minority groups actively participate in every other aspect of London's public life.

POLICIES FOR SUSTAINABILITY

As the studies of London's environmental and community health become more comprehensive, politicians and administrators, in consultation with the myriad of organisations who need to be involved, should be able to formulate longer lasting environmental policies. These should be related consistently and constructively to the goals of economic growth and the improvement of society, which are generally considered to be the primary objectives of urban development (Chapter 5).

Environmental objectives, which have to last for longer periods than most economic and social policies, need to be relevant up to the end of this century and beyond. Although this is only a short passage in London's long history (Chapter 4), it is a period in which the global climate is changing faster than at any time since the end of the ice age; when London lay at the southern edge of the ice sheet and the path of the Thames was finally settled (Trueman 1972). It will be essential to monitor all the environmental human and social and economic consequences associated with climate trends, and at the same time make use of the steady improvements in the computer based predictions of climate and environmental change. Doubtless it will be necessary to modify the policies as the data and the predictions change. Such environmental predictions are uncertain enough over rural areas and over oceans, but they are even more so for urban areas whose environments and physical structure are determined by changes in the economy and society (Hunt 2004). These changes can only be predicted very approximately using long term trends, such as national or worldwide growth patterns of urban areas. There are strong incentives to improve computer models for predicting social and economic developments (e.g. Lautso 2004). They need to become more detailed and more accurate as the rate of change quickens, when the consequences of errors in forecasting and planning become even greater.

Devising a strategy for sustainable development requires choosing how best to move away from the current practices where far too much waste is generated, and energy and transportation systems operate well below optimum in terms of efficiency and discharge of pollution (see for example London Sustainable Development Commission 2003). However there are no generally agreed strategies (or appropriate catch phrases) for the best way to make this transition. Only the broadest directions of policy are indicated by the concept of sustainable development. With such ambiguity in the concept, it is not surprising that there is a low level of understanding about how the public should change their 'life styles', in order to achieve sustainability. Furthermore differing views are held by parliamentarians and government departments as to what sustainable development strategy should mean. In broad terms does this mean, for example, rapid economic regeneration, followed by environmental remediation, or should the regeneration be restrained by ensuring environmental controls and more equitable and inclusive social policies? Or is it possible to avoid such dilemmas by devising strategies that meet both these sets of objectives? For organisations and communities to find these 'win-win' options requires motivation, ingenuity and dogged determination. Public bodies need to appreciate that there are benefits in associating politicians with such successes, as well as encouraging them to take the risks involved in attempting to meet demanding and often contradictory objectives (Chapter 19).

In any strategy of moving a city towards the goal of sustainable development certain problems have to be dealt with more urgently than others. Some environmental remedies are particularly controversial because they affect so many people in their daily lives. In London, as well as dealing with waste, the other urgent actions will be those to reduce energy use, through policies on transport, energy conservation, and reduce greenhouse gas emission by emphasising of renewable and other non-fossil fuel energy. This is necessary for UK to meet its declared international commitments declared at Rio in 1992 and Kyoto in 1997 (HMG 2003). Other aspects on the environment have long term consequences on the community even if they generally receive less media and political interest, such as environmental health, biodiversity, pollution and security. All these issues will be affected by how rapidly London expands and in what ways this occurs (Chapter 12).

In particular will London's economic growth continue to add to the 'waste mountain', recently castigated by the European Environment Commissioner? Current policies of waste disposal are being pushed to the limit, with their reliance on clean incineration (which produces as a by-product the energy requirements of the surrounding districts (Chapter 5)), and on finding new sites for land fill. Building often takes place on the sites that are already full. All these solutions involve difficult environmental decisions which are politically sensitive within and outside London. Ideally in future the amount of solid waste for disposal will be substantially reduced (perhaps approaching continental levels) by people consuming less solid matter (e.g. paper) and recycling. Improved clean technologies for reprocessing are equally essential for the success of this strategy (Chapter 5).

The use of materials in a sustainable community also has to be considered in terms of how their original production may be affecting the global environment. Therefore the types of natural material used, especially from developing countries but also from the UK have to be monitored (Chapter 5). This is to ensure that they are not extracted from ecologically sensitive regions that need conserving to preserve biodiversity, natural habitats or natural features. Coastal zones in particular produce many products consumed in urban areas, including oil and gas, and also provide holiday destinations for tourists from urban centres. But their ecology is now at great risk (ACOPS-10C 2004).

Since London's transport system is vital to the city's economy and greatly affects its environment, policies about the system's future are integral to the whole strategy of sustainable development. However deciding on sustainable transport objectives is highly controversial. Consultations about transport planning generally involve more representative groups than any other issue. Politicians who might not care greatly about environmental problems are always alert to anything happening in this policy area, as are business, trade union and environmental groups and local communities. London has seen recently large numbers of residents coming out on cold evenings to public meetings on traffic regulations, a sure sign of popular concern. But, as older residents will recall, these events have raised nothing like the passions expressed in the 1960s to defeat the proposals for a 'box' network of elevated motorways like those of Tokyo or Toronto. There was a great fear of their damaging impact on the physical and visual environment, especially in the historic centre. By contrast the current range of options being considered in consultations and political debate is quite small, lying between more or less public transport and use of road space by private cars. All policy options now assume that road traffic moves on the existing streets pattern. The greatest recent transport controversy concerns differing views on public and private funding schemes for investment in the system and on the arrangements for their management (Chapters 14, 17, 18). Perhaps this reflects a more economically sophisticated electorate. A supporter's view of these changes is that private sector involvement ensures that the long term viability and maintenance of the systems can be carefully considered and exposed to public scrutiny, something that did not happen when it was run wholly in the public sector and investment decisions were taken expediently on an annual basis. Chapter 17 takes another view! As so often in London's history politics is also drama; this topic is the subject of David Hare's highly charged play about transport accidents recently staged at the National Theatre.

The largest investment planned for the future transport system will be the fast underground 'CrossRail link', which will connect, with minimum environmental impact, the new housing and commercial developments in the eastern Thames Gateway (Chapter 2), right across to London's main airport in the West. It will also link up with the central terminus of the fast European rail network at King's Cross. The growth of London physically and economically is predicated on the growth of both surface and air transport. The substantial local and global environmental impacts of the airports are discussed in Chapter 14, but other risks also have to be weighed in the balance. With winds blowing from the west, aircraft will continue to fly low over London as they approach Heathrow airport. Economics dictates that the size of the aircraft will increase. But it is assumed that the risk of a crash is very small, even though major cities such as Amsterdam, New York and outer London airports have experienced aircraft disasters in the past. A vital element of a sustainable city is that it should be prepared for any eventuality, and in London all such disasters whatever their cause are regularly rehearsed by all the public services of national and local government (In Chapter 13 a Londoners' safety charter is proposed for showing how the principal hazards are dealt with).

Before roads and railways the East–West link for London was the Thames. But progressively its use for transport of people and goods has declined, though solid waste is still most conveniently shipped down river this way. Although there are advocates of restoring a return to river transport with fast surface ferries, there seems to be no economic case for significant shipping of goods into the centre of London (Chapter 12). In some cities this still happens. In Houston Texas, oil tankers come up its ship channel to unload at oil refineries within a few kilometres of the city centre. This is not a sustainable policy since the consequences of worse air pollution endangering public health outweigh any short term economic advantages.

Reducing the global impact of the world cities situated in developed countries should be technically and politically easier than in developing countries where some cities' populations may double in the next 50 years. By contrast London's population is growing slowly and by 11% over the next 20 years (DETR 2000). The objectives set out at the UN Environment Summit at Rio in 1992 for mitigating human effects on climate change and on the destruction of biodiversity were demanding, but, as it now appears, not demanding enough. These require cities to reduce substantially their emissions of greenhouse gases (GHG), and their consumption of environmentally sensitive material resources. The original targets of reaching by 2010 a 20% reduction of the 1990 levels of GHG emissions were set so as not to be too daunting. In reality these need to be further reduced, in total by 60% (RCEP 2000) if the quantity of GHG in the atmosphere will be stabilised at levels that will not seriously disturb the world's climate and its effects on the whole biosphere. A number of political, technical and financial measures being taken in London by the Mayor and the GLA, by central government, by business and by local communities, are showing that energy can be generated, utilised and also distributed much more effectively than at present (Chapter 5). The borough of Woking just outside London has brought all these aspect together to reduce by 40% the energy used by the Council's own buildings and operations. The carbon dioxide emissions are down by 70%. With London lying near the windy stretches of the Thames estuary, more of its electricity will doubtless come from wind power and other sources of renewable energy such as bio-fuels, which can be produced economically in the UK. London's proximity to the central government and financial markets are being used to ensure that the investment in these and other systems receive the appropriate finance and also tax incentives. As a global financial centre trading in carbon, London is also contributing to the more efficient use of energy world wide.

GOVERNANCE AND SUSTAINABILITY

The physical face of London and all the other aspects of London's environment are far from being the haphazard result of social and natural forces. They have been greatly influenced by many deliberate decisions of national and local governments and their agents over centuries, and also by how the patterns of housing, commerce and industry and transport have changed over time. The layout of streets and buildings were not suddenly transformed as in other major cities, by governmental convulsions as in Paris or a disastrous earthquake as in Lisbon. Rather they have evolved slowly, even though London has not been immune from fire, disease, and war. Just as significant in determining these and many other aspects of the environment were the responses of governmental bodies to popular and economic demands, particularly for new housing, improved health, sanitation, and transportation within and into the city. Inner London's street plans largely survived the disasters of the 17th century and the wars of the 20th century despite revolutionary proposals for their transformation drawn up by Sir Christopher Wren in the 1660s and Patrick Abercrombie during the 1940s to rebuild parts of the city completely. In fact pressure from local wards and parishes and public consultation only permitted some limited alterations to the basic street plan (Aykroyd 2001; T. Hunt 2004). No motorways were built in London, though large housing estates were (Chapter 15). Nevertheless some elements of those ambitious plans were enacted and with spectacular results; Wren's new St. Paul's Cathedral was erected, and in the last 50 years the South Bank of the Thames was fully opened as a public space for walkways, theatres and concert halls.

Despite its resistance to revolutionary plans, London has acceded to continual changes in the functional and visible forms of its governmental organisation. Some of these arose from the city's growth, and some from environmental pressures when they were too much for the existing organisations to deal with. (By contrast in Rome the street drains are still marked 'SPQR', 'the Senate and people of Rome'.) These pressures brought about the establishment of new local government boroughs for the expanding areas of London and new bodies for performing new functions, such as the Metropolitan Board of Works in the 1850s set up to provide London with clean water. Over the next hundred years environmental responsibilities were taken on by other public bodies, the main new ones being the London County Council (LCC) and London Transport. The great playwright and Fabian Socialist George Bernard Shaw believed these bodies were so worthwhile that he successfully stood for election as a Borough Councillor in St. Pancras.

Leading academics were also attracted to these bodies to work as public officials. The LCC established the importance of science and technical education (e.g. Allen 1933), a tradition continued by many public bodies in London. This has been further strengthened by the growing collaboration between local government, universities and government agencies, including, as we see in this book, research projects and information exchange on environmental and sustainability issues.

These new municipal bodies attempted to use their powers to ensure that infrastructure and housing developments made effective use of space and resources. In recent years the central government has tended to reduce these powers for strategic reasons. In some of the developments whole areas were transformed, for example in the suburbs and in central boroughs following war damage (Chapter 15). But mostly the earlier street plans and open spaces were preserved. During this time rail, road and canal transport networks expanded and then contracted. Meanwhile the locations and activities of industry and commerce were changing equally rapidly. The general planning strategy of these bodies, which still influences London today, was aimed to provide a high proportion of families with low rent housing situated in the central London Boroughs. They could use public transport to get to work on the London Transport system (Chapter 17), which was established as a successful public enterprise. It also took a broad view of its business when it established its own and London's visual identity associated with red buses and the ubiquitous underground sign. Leading designers were employed to create a new genre of advertisements on posters to encourage Londoners and visitors to make maximum use of their public transport system. For holidays and at week-ends they were enticed to benefit from the healthy air in the open parklands and to explore the countryside and historic sites in the suburbs. London's transport system has always had a considerable interest in serving the commuters from the outer suburbs, an interest that will continue even as London's policies for a strong central development receive a new breath of life with plans for the 'compact city' (Chapter 2, 19).

London's achievements in 19th century municipal city government were widely admired and copied, for example in Prussia. But in the 20th century London began to take ideas from continental cities, where, especially in Germany, they have provided many of the best examples of how local government can deal with environmental policy, waste, noise and public transport. Barcelona has brilliantly exploited the use of valuable urban space for housing by building very much more intensively than in London (Chapter 2).

In the past 30 years London's local government bodies have been reformed several times as the views of UK central government and business interests have veered between limiting and expanding the roles of local communities in economic development, planning, transport and environmental responsibility.

To expedite the regeneration of the economically decaying dockland areas of London, a new Docklands Development Corporation was set up in 1985 with stronger planning and development powers than those of the local London boroughs and the city of London (Chapter 18). This ensured strong coordination of the multi-billion pound investment in Canary Wharf, North Greenwich and other huge commercial and housing projects. The necessary extensions of the surface and underground rail networks were funded by central government. The timely completion, financial viability and general success of these projects have been in part a result of their particular governance and regulations. This approach has given London the confidence to make further massive extensions of these projects further east along the banks of the Thames Estuary.

When Parliament came to set up a new structure of Government for London in 1999 with a directly elected Mayor and a Greater London Assembly, the new body was given clear responsibilities for the environmental strategy which unusually were set out in great detail 'on the face of the Bill'. The progress made in establishing and beginning to implement these objectives is reviewed here in Chapters 5 and 19. The Mayor is obliged to take the lead in collaborating with many other bodies having day-to-day responsibilities for the environment and sustainable development. As a result of privatisation policies many of these bodies are no longer publicly owned, though they may work for and are regulated by public agencies, such as those responsible for water supply and solid waste removal. Another dimension of governance has sprung up with community groups and private companies introducing and disseminating new techniques for safeguarding the environment. These in turn are involving ever more organisations in London in sustainable development initiatives (see Appendix and Chapter 11).

A striking aspect of these developments in governance has been that environmental decisions in London, as in many other world cities, are now being decided more openly than for most other areas of policy, and often at a local level. This is essential in London when there are so many different bodies with overlapping responsibilities scattered across the city. Just one example makes the point; there are more than 100 bodies managing parks and gardens in London, whereas there are about two in Paris. Only with open information can so many organisations work effectively together through informal cooperation, which is the usual way for British public bodies to work. This kind of bureaucratic self-help and popular involvement is essential since only rather broad directions and limited funds are usually provided by national or local government.

Openness pays off when it is combined with vigorous programmes of regular information; also communities need to be involved in every aspect of these developments (Chapter 15, Jones 2004, House of Lords 2004). This encourages people to undertake energy conservation measures in their own homes and in their use of transport, while generally support-ing civic and commercial programmes of sustainable development, for example by buying 'green' energy. Greater understanding and effective application of new clean technologies should be an integral part of this community based approach. Since the global environmental impact of conurbations is disproportionately large, the choice for cities is stark. The whole population of the world will benefit if they can ensure that their future development is more sustainable (Chapter 16). The whole world will suffer if they fail.

As these issues progressively dominate every aspect of urban life, everyone including politicians, needs to understand them. We hope that this book will help!

REFERENCES

Acops-IOC (2004) Coastal zones in Sub-Sahara Africa. A Scientific Review of the Priority Issues Influencing Sustainability and Vulnerability in Coastal Communities. Intergovernmental Oceanic Commission, Paris.

Ackroyd, P. (2001) *London, a biography,* Viking, London.

Allen, W. (1933) Memoir of W. Garnett, Heffer, Cambridge.

Booth, C. (1903) Life and labour of the people of London.

Brundtland, G. (1987) 'Our common future'. World Commission on Environment and Development. Oxford University Press, Oxford.

Chandler, T. J. (1965) The Climate of London, Hutchinson, London.

Clout, H. (ed.) (2004) The Times London History Atlas, 2nd Edition. Times Books. London.

Dept. of Health (2001) Health effect of climate change in the UK. Dept. of Health, London.

DETR (2000) Our towns and cities: the future. CM 4911. DETR, London.

Goode, D. A. (2000) Cities as the key to sustainability. In: Where Next? Reflections on the human future. Ed. D. Poore. Royal Botanic Gardens Kew, London.

Hansard (2000) House of Lords debate on 'Long term problems caused by erosion'. 7 June. Parliament. London.

Her Majesty's Government (HMG) (2003) Energy White paper.

Her Majesty's Government (HMG) (1995) Environment Act.

Howard, L. (1833) Climate of London. Harvey and Dartxon.

House of Lords (2004) Climate change in the European Union, Report of the E.U. Subcommittee D. Stationery Office, London.

Hunt, J. C. R. (2004) How can cities mitigate and adapt to climate change? *Building Research & Information* **32**(1).

Hunt, J. C. R. (2005) Inland and coastal flooding: developments in prediction and prevention. *Phil. Trans. Royal Society* (in press).

Hunt, Tristram (2004) Building Jerusalem, Weidenfeld & Nicholson, London.

Jones, A. (2004) Evidence to House of Lords European Union sub-committee, February 25.

Klinenberg, E. (2002) Heat wave, a social autopsy of disaster in Chicago. University of Chicago Press, Chicago.

Lautso, K. (ed.) (2004) Propolis: Planning and Research of Policies for Land Use and Transport for Increasing Urban Sustainability, LT Consultants, Helsinki, Finland, 369pp.

London Sustainable Development Commission (2003) Consultation and sustainable development framework indicators. February. GLA, London.

Royal Commission on Environmental Pollution (2000) Energy — the changing climate, London.

Trueman, A. E. (1972) Geology and scenery in England and Wales (Revised by J. B. Whittow and J. R. Hardy), Penguin, London.

UKCIP (2002) see www.changingclimate.org.

UN (2001) Population statistics, United Nations, New York.

Chapter 2

London's Urban Renaissance

Richard Rogers

The British are famous for their cities, which they both love and hate. They gave to the world the idea and the practice of how cities can flourish and they pioneered how to deal with some of the worst environmental and social aspects of the industrial revolution in urban areas. Between the 1830s and 1860s about 70 cities were founded in Britain. The Prussians and many other nationalities came to see these extraordinary legal and administrative innovations, which they copied, developed and often improved in their own countries. Today 90% of the British live in cities. London, through effective action by government and business at every level, continues to prosper and attracts migrants from all over the world.

But London's future quality of life and cultural meaning is in danger because its whole environment is deteriorating and becoming notably worse than that those of other major European cities. The great monuments, parks and vibrant life cannot disguise the squalor of many streets and whole districts of London which are an affront to everyone whether visiting or living here. This situation is not only hugely damaging to the UK's commercial competitiveness, but the longer it is left, the more expensive it will be to rectify. The enormous and growing disparity of wealth, education, justice and health in its population makes the city progressively more socially dysfunctional, whose knock-on economic consequences are only beginning to be understood.

What are the main physical and political principles to guide the development and government of a well designed, well run and sustainable city? Firstly there need to be strong policies about connected planning, design and transportation; the overall layout of the successful city needs to be compact with a small number of separate centres — the 'polycentric form' — so that local hubs of public transportation, and other elements of the urban infrastructure can operate more efficiently. The value of parks, open spaces and waterways to a city is also much greater when they are connected in green corridors, as German cities have shown. It is essential that these corridors extend into the poorer areas where industry was concentrated in the 19th century.

For reasons of cost, social cohesion and to preserve the countryside, urban housing densities should be increased to approach those found in major continental cities, where for example as in Barcelona there are 400 dwellings/hectare which are mostly apartments. This compares with the current plans for about 75 dwellings/hectare in London where most people live in houses. These densities are far too low if the Government's targets for new housing in London are to be met. Provided the designs are sympathetic in the British context, continental densities of housing should be practical and be accepted here, especially in the huge Thames Gateway development in East London, which will become UK's gateway to Europe. This is the cornerstone of the Greater London Authority's plans to house 700 000 in London — the expected population increase over the next 15 years (roughly the population of Leeds) — mainly on 'brown field' sites so as not to encroach on the greenbelt and the ecological reserves around London.

In terms of architecture and urban planning, the souls of our city centres are fragmented because of an essentially ad hoc approach to redevelopment. We are close to a meltdown condition exacerbated by the fact that there has been no significant investment in planning or urban development skills for decades. And this despite the fact that — after Bangladesh and Holland — Britain is the world's most densely populated country. Holland channelled billions of guilders into building and infrastructure projects in the 1970s and 1980s to keep ahead of the urban development game — and used largely untried but highly talented younger architects to do it. Britain's urban sin of omission has had an often devastating effect on both public and private sector architecture and infrastructure, especially the architecture of smaller urban spaces, parks and squares, where people watch people. One thinks of Barcelona, Paris, Copenhagen. This lack of any decent built environment has a devastating effect on our sense of identity and culture. If our towns and cities are vague urban smudges rather than places with a properly defined and organic presence, we might ask ourselves: are we British, or European, or just bit-players in an unravelling soap called 'Grungeville'?

The transportation planning systems of major cities, even in the UK, differ more than any other aspect of urban planning. Central London's transportation is more sustainable than most other urban centres in the UK in that about 81% of people use public transport to go to work, whereas elsewhere it is generally below 20%. But the nature of the 15% private transportation in London, which increased from about 7% in the 1960s, is causing air pollution damage to people's health and to the fabric of buildings. The use of cars in outer London is significantly greater, which is the reason why air quality there is often worse than in central London. This growth of car traffic despite some improvements in technology feeds a vicious circle, encouraging fear of public spaces and the belief that only private spaces and transport are safe, especially for children. This rich/poor division of city life is seen only too clearly in Los Angeles and Phoenix. London needs safe public transport arrangements for journeys to schools. Cycling and walking are on the increase with better cycle lanes, wider pavements, and longer phasing of the traffic lights, but London contrasts very unfavourably with Copenhagen where cycling is about one third of the traffic, by comparison with London where it is about 1.5%. The poor design and maintenance of streets are one of the main manifestations of urban squalor and a deterrent to more use by cyclists and pedestrians. Another very 'green' form of transport is the use of waterways; there should be much greater use of the Thames by small boats. This too requires planning and resources, and determination to 'work round' conflicting demands of different uses on and along the river.

The second set of principles is concerned with the ecology of urban areas and the wider issues of environmental degradation, reducing environmental risk, loss of biodiversity and maintaining sustainability. Although these issues are considered in more detail in other chapters, it is essential to note the vital roles of architects and planners. The buildings, parks and cities they design substantially affect the environment at all scales. They are also affected by the environment, including long term climate changes on the scale of the whole Earth. More needs to be done to understand these interconnections, especially since the worst effects of climate change and polluted atmospheres can be mitigated by better design involving low energy buildings, less traffic with cleaner engines, and the careful preservation of open spaces.

London is monitoring its ecological footprint and taking the first steps to mitigate climate change effects, through the efforts of many organisations. But more consciousness raising, commitment and resources are needed.

Finally the people living in a city will only be able to ensure its long term economic and social success and sustainability if they choose to be governed democratically both locally and city wide. They also need to support a diversity of organisations that contribute to the well being of the community. Most social groups, no matter how egregious, can be included in the urban society. London has passed through a period of semi-direct rule to having, since 2000, its own overarching but directly elected Greater London Authority, working with a national government that has focussed more on urban regeneration than most of its antecedents. The Urban Task Force, which I chaired with 14 colleagues, reported in 1999 and some but not all of its recommendations were included in the White Paper policy document 'Delivery and Urban Renaissance'. Regrettably it did not provide real financial incentives to clean up and develop brown field sites. At the same time the main element of the London's governance and social organisation have continued as before with a strong city corporation and strong London Boroughs, both of which are introducing greener policies. Many new voluntary groups have sprung up with new ideas and solutions and the commitment to involve more Londoners than ever before in key environmental and sustainability decisions. However none of these developments in governance and local participation will lead to a better London without a deeper and prolonged involvement by architects and planners in the sustainability of their projects, and by their pressure for changes in how projects are initiated, funded and delivered. 'Green lipstick' is not enough.

REFERENCES

DETR (2000) Our town and cities, the future. Delivery on Urban Renaissance. Stationery Office, CM 4911.

Rogers, R. (1999) Towards an Urban Renaissance, Urban Task Force, E & FN spon.

Rogers, R. (2002) Observer July 21, *Delivery and Urban Renaissance.*

Chapter 3

Sustainability of London's Environment and the World Context

Michael Meacher

INTRODUCTION

The absorbing overview of London by Richard Rogers in Chapter 2 links to my concern about its overall sustainability. This book is based on the first conference to be held on London's environment for a long time. It was an impressive one. The timing was pretty good too — 50 years on from the great smog in London in December 1952, and immediately following the World Summit on Sustainable Development (WSSD) in Johannesburg in August/September 2002.

First the context: the environment is one and indivisible. If we think that London has some difficult issues to tackle, let's just stop and think about some of the problems facing the developing world, that were the chief concern at the Johannesburg Summit. This was held ten years after the UN Earth Summit at Rio de Janeiro in 1992. But still, 200 years after industrialisation more than 1.2 billion people, one in every 5 on this planet, live on less than one dollar a day, and about 2 billion people have no access to electricity. At least 1.1 billion people lack access to safe drinking water and, what I think is the most shocking indictment of all — over 2 million children die every year from drinking contaminated water and diarrhoea related diseases. That's about 6 thousand per day which is twice as many people as were killed in the terrorist attack on the Twin Towers, New York in September 2001. And these children die in these numbers every day.

JOHANNESBURG

So after all the rubbishing and trivialisation in the media was the Summit at Johannesburg (WSSD) worth it? The answer is 'yes'. We now have a clear global target to halve the number of people without access to freshwater and basic sanitation by 2015. We also agreed on a number of actions; a commitment to a national programme to increase the use of renewable energy; action to protect biodiversity and natural resources, particularly action to conserve fish stocks (60% of which are currently being fished to extinction) and the development of marine protected areas where little or no fishing is permitted; and to minimise the significant adverse effects of chemicals on human health and the environment by 2020.

These are just a few examples. But one of the most significant outcomes for developed countries, and particularly the UK, will be the development of a 10 year framework of programmes on sustainable production and consumption. What this means is a fundamental shift towards production and consumption that is within the carrying capacity of global ecosystems, which is certainly not the case at the moment, with the over-exploitation, over-pollution and over-utilisation of resources in current production and consumption patterns in the West.

That means a huge campaign to raise the public awareness of these issues, which is still far too low. There is still far too little understanding of the inordinate and relentless pressure we are placing on planetary, and particularly urban, resources.

LONDON

So after all the millions of words in Johannesburg, what are we going to do here in the UK and London? Well, London is not only our capital city, it is also a financial, commercial and cultural centre of world importance. However, its sheer size and the activity it generates puts considerable pressure on the environment in which it operates. For example, the London City Limits project, which has just produced a report on London's ecological footprint, indicates that each year London produces enough waste to fill the Royal Albert Hall over 200 times. That is not sustainable in the long term. Also as a city, London consumes more energy than Ireland and about the same as for Portugal and Greece.

The draft London Plan which sets out the Mayor's vision to develop London as an exemplary, sustainable world city predicts that the population of London will reach 8.1 million by 2016, about 700 000 (equivalent to the population of Leeds) more than today. The challenge will be to see whether growth on that scale can be accommodated without having a detrimental effect on London's environment and people's quality of life.

Earlier this year the Mayor of London, Ken Livingstone, established the London Sustainable Development Commission and I was pleased to see that one of the Commission's first priorities will be to produce a Sustainable Development Framework for London with objectives, targets and indicators. I hope and believe the Commission will be able to make an important contribution towards making London a more sustainable city.

Three aspects of London's environment that present particular challenges are air quality, climate change and waste.

AIR QUALITY

As I stated earlier, this year is the 50th anniversary of the great London smogs of 1952. Air quality in London has improved considerably since then. More recently, levels of most pollutants have fallen considerably through measures to cut emissions from industry and traffic. But the latest health evidence shows that we cannot afford to be complacent. That is why in August 2002 as Minister of the Environment I announced a significant strengthening of our air quality targets for particles and other important air pollutants.

The targets for air quality for London differ from the rest of the country as they take account of the markedly higher levels of air pollution in the capital. But the objectives we have set for London are equally as challenging as those proposed for the rest of the country. The level of reduction should be the same if not greater, and Londoners should experience similar, if not greater, improvements in health benefits as the rest of the population on the basis of present policies and measures. (DEFRA 2000)

CLIMATE CHANGE

Climate change is an increasing threat that could affect the whole of London's infrastructure and environment through rising temperatures and greater risk of flooding, leading to worsening air quality, damage to buildings and property, and disruption of the transport network.

In October 2002 I launched the London Climate Change Impacts study, which was commissioned by the London Climate Change Partnership (2002) — a broad range of key stakeholders in London. This was a particularly important piece of work, as the London report was the first UK impacts study with a distinctly urban focus, and one of the first studies to focus on the impacts of climate change on a major world city.

London also plays an important role in our national programme to reduce greenhouse gas emissions. For example, as part of our Climate Change Programme, we launched the world's first economy-wide greenhouse gas emissions trading scheme earlier this year. This scheme is designed to consolidate London's role as the future international centre for carbon trading. (House of Lords 2004)

WASTE

London produces 18 million tonnes of waste every year — a staggering amount which continues to grow. Overall, London recycles less than half of this waste. Only 8–9% of London's municipal waste is recycled and over 70% of it goes into landfill sites most of which are outside London. This is clearly unsustainable. The Landfill directive and the targets we have set in the National Waste Strategy (Strategy Unit 2002) require substantial reductions in the waste we landfill and big increases in recycling and composting.

The UK government has made £140 million available over the next two years (2002–2004) to help local authorities in England achieve their statutory recycling and composting targets. In London, I was pleased to award £21.3 million from this money to a partnership of the Mayor, the Association of London Government and London Waste Action to distribute to London boroughs on a strategic basis. And in September 2002 the Mayor published the draft London waste strategy.

CONCLUSION

So, in conclusion, London's environment is of the greatest importance to government, especially at present. The world summit presents us with a huge challenge, both in the UK and particularly here in London. The evidence is overwhelming that we cannot go on as we are, in so many aspects of our economy, our society and our culture. The question is, are we ready?

REFERENCES

London Climate Change Partnership (2002) London's Warming, The Impacts of Climate Change on London, Greater London Authority, London.

DEFRA (2000) Airquality Strategy Cm 4548, Stationery Office, London.
House of Lords European Union Committee (2004).
The EU and Climate Change HL Paper 179, Stationery Office, London.
Strategy Unit of UK. Gov't (2002) Waste not, want not www.piu.gov.uk/2002/ waste/report/index.html

Chapter 4

Foundations of a World City — London's Historic Environment and its Future

Taryn J P Nixon

London is the creation of layer upon layer of complex and fascinating interactions between people and place over thousands of years. People have been living, visiting and working on this very spot for over 500 000 years, each generation bringing its own entrepreneurs and architects of progress, each government and local authority with new ambitions and wielding new controls. The World City that is London is the result of layers of ideas, some of them good, some of them enduring, some of them neither.

It is therefore no surprise that we have an enormous thirst for information about the history of London. A survey by the British Tourist Authority showed that 54% of overseas visitors rated historic buildings as one of the reasons for visiting London (source BTA 1995, in English Heritage 2002). Overseas visitors, according to the Office for National Statistics, spent 13 150 million nights in London in 2000, which makes it by far the most visited destination in the UK for foreign visitors. Tourism accounts for 4.9% of the Gross Domestic Product (GDP) of England and about 7.6% of employment in the UK, and cultural tourism in particular is of enormous value to the economy of London (ibid).

The growth in popularity of history is reflected in a surge in the number of national and regional television programmes about archaeology and history. What is behind this fascination? It is more than simply the lure of a good detective story (although that certainly helps); it appears to be much deeper, tapping into the relevance of the past to our future. Our origins and roots are vital to creating a sense of identity and pride not only in our inheritance but, crucially, in our plans and aspirations for the future.

The past, in other words, has a pivotal relevance to the future. Not only does history teach us lessons and inspire our dreams, but, like the natural environment, the historic environment is vital to our sustained social and economic growth.

A SENSE OF PLACE: LONDON PAST

To learn those lessons and to use our past to benefit our future requires an understanding of the history of London's environment. Rather than attempt the impossible by summarising half a million years of human activity in a few paragraphs, this chapter touches lightly on different times and themes in London's past, drawing on the research questions that most occupy London's archaeologists at present, as set out in the Research Framework for London Archaeology 2002 (Nixon *et al.* 2003). These questions are also considered from the perspective of the current Government's 15 headline indicators for sustainable development (HM Government 2002), in particular:

- environmental protection (climate change, air quality, road traffic, wildlife, waste, river water quality and land use),
- social progress (poverty and social exclusion, education, health, housing and crime) and
- economic growth (economic output, investment and employment).

How successful, one wonders, have past Londoner's been in safeguarding the city's environment? There has been human occupation in London for some 500000 years from the Lower Palaeolithic through to the present day. In the Early Holocene, around 10000 years ago, the Thames ran to the south of the modern course. Evidence from excavations in Southwark and Lambeth suggests the river was braided, made up of many channels and dotted with sandy islands or 'eyots'. After the end of the last glaciation around 10000 years ago, at the start of the Holocene, London's landscape was one of mixed oak woodland with forest-dwelling animals like red deer abundant. Average summer temperatures were several degrees higher than today (Evans 1975), probably comparable to those predicted for 2080; when one in three summers are estimated to be hotter than the unusually hot summer of 1995, which was 3.4 degrees Celsius warmer than average (Hulme *et al.* 2002). (See Fig. 1).

Fig. 1 As the boroughs of Southwark and Lambeth would have looked around 7000 cal BC (K. Singh, MoLAS).

Subsequent deforestation was thought largely to represent the start of early agriculture: deliberate clearance of woodland would have opened up the hunting grounds and created arable fields. Recent archaeological excavations in Southwark and Lambeth also show a steady decline in elm and lime trees from disease; however the decline in limes may also indicate increasingly wet conditions (Sidell *et al.* 2002). Was this a consequence, even then, of failing environmental protection? The shift from hunting and gathering to farming was a major change for Londoners and the archaeological record begins to show dramatic evidence for social progress. From the sedentary, agricultural patterns of the Neolithic some 6000 years ago, London's landscapes continue to evolve and to be modified by people, with the arrival of major, monumental and ceremonial landscapes of the Bronze Age around 4000 years ago, and the settlements of Iron Age London revolving, in the main, around an agricultural economy (see Fig. 2).

When the Romans arrived in the 40s AD, they would have found a varied landscape of cultivated ground, grassland, forest and wetlands. Roman *Londinium,* of course, was an 'implant' on the native landscape. The Romans deliberately built a Roman city, with a road network, key public buildings like an amphitheatre, a market place and administrative centre, a fort, and so radically altered the landscape. Big public buildings were built on the banks of the Thames, with highly visible and often imposing architecture that would have proclaimed not only the importance of the city but also by implication its status (Millett 1990). They reclaimed land, consolidated the north bank of the Thames and built the first bridge across the Thames. In terms of social inclusion this process of 'Romanisation' was, in fact, very effective cultural interaction (Woolf 1998). To be a 'Roman Londoner' was to live in a complex and diverse society, one which combined a cultural affinity towards the legacy of Rome — through language, literature, law, architecture and art — with the varied cultural traditions of the native Britons (Bradley 1994). This was work that would be worthy of a Regional Development Agency today, with a vision for urban renaissance and regeneration. The archaeological evidence for social growth throughout the Roman period paints a picture of tremendous diversity in religious belief, with an abundance of coexisting religions and cults seen through various temple complexes, a Mithraeum, a possible cathedral, and so on. Indeed, it may have been this curious blend of coexisting belief systems that gave its unique character to London and to its surrounding countryside.

Fig. 2 There is evidence for extensive deforestation by around 2500 cal BC, followed by further clearance of remnant woodland isolated copses (after Merriman 1990, Prehistoric London, London).

We recently learned much more about the design and management of Roman water supply, with the discovery of three large wells at 30 Gresham Street, City of London, and the associated wrought iron and wooden water lifting mechanisms. Massive wheels, apparently powered by slaves or animals, would have been capable of raising and supplying water to a population of 10000. Was this evidence of centralised investment in the infrastructure of the city? Certainly, drainage was equally well planned, at least in relation to the public buildings. The surviving storm drain that probably ran from the basilica and forum (beneath today's Leadenhall Market) down towards the Thames, is further evidence of concerted water management.

Unsurprisingly, economic growth was rapid; massive timber quays and warehouses were constructed to service the major Roman port. By 60 AD London had become the biggest town in the new province and a thriving cosmopolitan immigrant settlement. Although it would seem that the investment of the early years was not consistently sustained, and London in the 2nd century may not have been especially healthy, with its smelly roadside ditches and silted up streams. It was very likely seen as a place of opportunity but the archaeological evidence tells us it was environmentally unregulated and unenhanced. The Romans were by no means 'green' — but they did, however, continue to manage the surrounding woodland, agricultural supply and extractive industries. The economic drive to manage the relationship between city and rural hinterland was powerful: then, as now, London was a major consumer of raw materials, from grain and basic foodstuffs, timber and fuels.

After the collapse of Roman Britain, and the reversion of much of London's environment to mixed deciduous forest, the powerful influence of the Thames is once again thrown into stark relief. The tidal Thames was apparently the single biggest influence on Saxon settlement: it was not only an important food source, but also a transport link providing access to markets, routes for settlement and migration, and indeed it was a barrier for defence. The main Saxon town, Lundenwic, was sited at Covent Garden; cemeteries and settlements were concentrated along the Thames and its tributaries. The basis of the Saxon economy was clearly agricultural, but evidence is sparse and we know virtually nothing about the farms that supplied London, or the rural settlements of the region, or how rural and urban diets compared. We can glimpse social status through the grave goods from Early Saxon cemeteries and through the refuse from Saxon houses. Once again, there are hints at a stark difference between town and country, with a lack of luxury imports into the countryside.

The basis of medieval London, as in the periods before and since, was its role in importing and exporting, with the city dominating the region. Medieval London was probably the single largest concentration of industrial production in England, with over 100 crafts within the city walls, such as metalworking, bell-making, textile manufacture, horn preparation, and pewter-making. The City population between 1100 and 1300 saw a rise from 25 000 to perhaps 60 000 or 80 000. What we don't have is the equivalent estimate for the modern Greater London area. We do, though have information from both archaeological and documentary sources, including an assemblage of over 17 000 human skeletons excavated from across the region, spanning the period from 1100–1500. This is a resource of unparalleled importance, with the potential to explain the population's changing demography, health, and levels of personal hygiene (see Fig. 3).

London's inexorable growth over the last 500 years is well known. The population of Greater London rose from an estimated 120 000 in 1550 AD (after Finlay and Shearer 1986) to just over a million in 1801 (Beier and Finlay 1986). During the 17th century the central conurbation was divided into three distinctive areas: Westminster — a political and social-area — the City and Fleet Street — a commercial, financial and legal district, and the East End — associated mainly with trade and industrial activities. From the 16th century, refugees, minorities and non-conformists were settling in London and establishing their own places of worship; these were the roots of today's urban multiculturalism. Opportunities for social interaction were many and varied, for example there were theatres and bear baiting on the South Bank, situated on cheaper land where some of the smellier industries, such as fish farming and tanning, were also located (see Fig. 4).

In the 17th and 18th centuries, amid fierce economic growth, society was heavily influenced by the import of new commodities. Beverages such as coffee, tea, chocolate, punch and gin arrived, and coffee houses proliferated; daily life was underpinned by an elaborate infrastructure of public and civic buildings (including town halls, hospitals, almshouses, prisons, schools and workhouses), markets, transport, water supply, waste disposal, communications, and other services such as gas and electricity. Canals and, in the 19th Century, railways played a key role in supply and production, and, rather like the Roman roadside settlements, post-medieval

Fig. 3 Thorney Island, one of the sandy eyots that existed with in the Thames in prehisotric times, on which Westminstex Abbey and the Palace of Westminster were built in Saxon times (A Chopping, MoLAS).

coaching inns played an integral part in the nation's transport network. There were changes in water supply and waste disposal that had a major impact on London's rivers and hence on public health and the river and upper estuary fisheries. The docks and shipyards of the East End thrived.

Fig. 4 Modern London bridge with a reconstruction after Peter Jackson of London bridge in C1600 superimposed (A Choppping, MoLAS).

The growth and importance of market gardens also had a crucial role in provisioning the population; we know little about how they worked but we do still have surviving fragments of ancient woodland and open country preserved in parks, gardens and graveyards throughout London. Traffic congestion was by now a real problem.

Of all the re-current themes in London's rich and complex history, it is perhaps the multi-culturalism and diversity of Londoners that stands out most. History enables us to identify the different flavours and symbols of different parts of London and also shows us how the success of the city as a whole is inextricably linked to the way different areas, and different communities have sustained it. London has been a world city at many times, and certainly it is the world city it is today because of centuries of interaction between people and place. In striving for sustainable development in the 21st century, therefore, we would do well to draw lessons from the successes and failures of our past, in the balance between town and country. When we think of London we need to look much further than the urban conurbation: we need to look at the balances, the links and the relationship between all the urban villages and communities of London and its modern-day hinterland.

A SENSE OF PLACE: LONDON'S HISTORIC ENVIRONMENT TODAY

London today is a World City, with only perhaps New York and Tokyo to rival its business functions. It is the seat of national government and the financial capital of Europe, as well as being a major business and trading centre and the headquarters of choice for many national and international companies. It is still (or perhaps more accurately, once again) the most diverse city in the world: a third of London's population of 7.2 million people belong to a myriad of ethnic groups. Its built environment, and its parks and open spaces, are the product of at least two millennia of growth. Its street patterns and monuments make it instantly recognisable as a richly historic city, and this has made it one of the world's top urban tourist destinations. London's historic environment today comprises of 600 protected London Squares; 142 historic parks and gardens; 3 World Heritage Sites; 40 000 listed historic buildings; and 881 conservation areas (ibid).

Above all London is a dynamic city. The historic environment is subjected to change and just as London today is the result of complex interactions between people and place, it must continue to change. London's population is expected to grow by 700 000 in the next 15 years, and the draft London Plan cites the need for an additional 300 000 new homes for people living and working in London (Mayor of London 2002).

The challenge for those concerned with the future of London's historic environment is to strike a balance between continuity and change. Since the Earth Summit in Rio in 1992 (IUCN 1992), promoting the need for national strategies for sustainable development (Agenda 21), the historic environment sector has itself matured. The Council of Europe has declared that 'In order to prosper, towns must continue to change and develop as they have always done in the past. This means that a balance must be struck between the desire to conserve the past and the need to renew for the future' (European Council 2000). The London Plan (Mayor of London 2002) covers the historic environment in a brief section on built heritage and views, and seeks to 'ensure that the special character of historic and archaeological assets is recognised and understood and that this forms the basis for their protection and the identification of opportunities for their enhancement.' All these statements converge in one area: they promote the need for an approach based on establishing values of all parts of the environment, and a determination to see and treat the historic environment as an asset, not constraint, which we can only applaud. They bring us to a definition of heritage conservation today which can actually be described as 'managing change in order to hand on what we value to future generations' (Clark 2003).

A SENSE OF IDENTITY: THE ROLE AND VALUE OF LONDON'S HISTORIC ENVIRONMENT IN ITS FUTURE

In the face of demands for housing, infrastructure, economic and social growth, those responsible for the heritage sector face issues considerably more complex than simply making a choice between what to preserve or protect and what to sacrifice to modernisation. Their wider role will be to harness all the qualities and socio-economic benefits that the historic dimension can bring to the future environment.

The final section of this chapter is devoted to further aspects of this theme, addressing the vital role that the historic environment plays in sustainable development. The historic environment is just as much a part of our world as the natural environment. However, if policy makers, strategists, and shapers of our future neighbourhoods are to sustain responsible growth, then they will want to create places where people truly wish to live, work and visit.

We have already considered how an understanding of the past provides people with a context and a framework for our daily lives and our decisions about tomorrow. Our aims and aspirations are moulded by what has gone before. What is less well acknowledged is that history, in its widest sense, provides individuals and groups of people with a deeply felt sense of identity. This plays a central and essential role in engendering pride of place. Although history focuses on London as a single entity, as individuals our main interest is in our own neighbourhoods, and perhaps our family origins. In London, the sense of discovery is made all the more exciting because so much of the city's early history is obscured: today's modern cityscape bears little resemblance to very ancient historic landscapes and whereas we might still make out, say, old street patterns and buildings, modern suburban development has superimposed new characters on old places. What we have today, however, in spite of urban sprawl, is a series of related and overlapping neighbourhoods. They have very distinctive characters of great importance to those who live and work in them. Local histories continue to captivate. It is at a local level that we can best engage with people and help them discover for themselves how adjacent neighbourhoods with superficially similar building styles may have widely different origins. For example these may be traceable either to a Roman staging post or to a medieval market town.

In this way, it is clear that the historic environment is, in fact, a potentially powerful and effective vehicle for social inclusion. So surely, we should be working doubly hard to use archaeology and history to illuminate the present, by demonstrating how modern London has grown out of the past... and how past landscapes will continue to mould the future.

The concept of inherent pride in one's environment has been explored recently by English Heritage in the document 'Power of Place — the future of the historic environment' (2000) and in the Government's response to that document 'Force for our Future' (DCMS/DTLR 2001). This research included a MORI poll on attitudes to the historic environment, to buried and unseen archaeological sites as well as to London's visible historic buildings and it is clear that most people accept that the historic environment has a social value.

Until recently, however, the economic value of the historic environment was far less clear. Work by the National Trust has placed an economic value on the rural historic environment, in four major area surveys in England. For instance in the South West of England, tourism has been found to contribute £2354 million to the regional economy. Significantly, some 40% of jobs created through tourism are directly dependent on a high quality environment. 'Quality of life' is clearly enhanced by the celebration of the historic environment. Consequently preserving archaeological and historic sites and landscapes has a real economic value. It is equally important to understand that the historic environment is part of the natural environment. In short, this work has demonstrated that the rural economy depends on the *quality* of the natural and historic built environment (National Trust 1999).

A number of methods have been used to determine economic values for heritage (for example Graham *et al.* 2000), such as hedonic pricing (where heritage is assumed to be capitalised in real estate prices), travel-cost pricing (where the costs of visiting a site will reflect the value placed upon it), contingent pricing (which requires individuals to place a value on heritage in a hypothetical contingency, such as the destruction of the monument or site) and 'Delphi' methods (where experts or stakeholders provide a view). For London, the newly established Historic Environment Forum is exploring heritage values for the nation's capital. This Forum is convened by English Heritage (London Region), with membership comprising the National Trust, the Greater London Authority, the Heritage Lottery Fund, the London Tourist Board, the Museum of London and others. One of its main aims will be to promote the real asset value of London's historic environment and realise its socio-economic value.

The challenge lies in making the link between archaeological sites, monuments and historic buildings, and contemporary townscapes, whether through physical conservation or functional conservation. The newly re-opened Roman amphitheatre is a fine example of physical conservation, where the Corporation of London has preserved and displayed the arena walls in a dramatic and evocative exhibition beneath their new Art Gallery building at Guildhall, in the City. There are countless other historic structures, incorporated into modern fabric or modern streetscapes. However, given that the discipline of archaeology is about the history of people, rather than things, it is important to distinguish between physical conservation and functional conservation. Most of our existing conservation policies aim to preserve the physical fabric: protecting historic structures. But physical conservation on its own tends to lead to empty, lifeless, open-air museums. Functional conservation, on the other hand, is about how space is used and offers the opportunity for people to have pride in their surroundings. This approach takes the history and traditions of a place and makes them the focus for the future. It can mean re-using an historic building or space or street pattern. But equally it can provide themes even without the physical preservation of bricks and mortar.

In other words, the historic environment's contribution to London's future might, above all, be in the spirit of a place — its 'genius loci'. Our history is undoubtedly a catalyst for regeneration, encouraging not only overseas but also domestic tourism and nurturing the pride of local communities in their neighbourhoods. For instance, one aim might be to take London's manufacturing traditions, which are now in decline, and integrate them into the character of neighbourhoods of the future. Indeed, many regeneration schemes do this by seeking to reconnect historic sites and landscapes to contemporary townscapes, in order to give the neighbourhood and its people a sense of identity. The architectural challenge, lies in drawing out the amenity and aesthetic values of historic assets, as is so evident, for example, in the heart of Southwark's Borough Market, the City's Leadenhall Market, or the new architecture of Shad Thames.

At present there is sometimes conflict between the aims of the historic environment sector and the aims of our city developers. Requirements for energy efficient buildings, for instance, conflict with the conservation needs of historic buildings; agricultural policies directly conflict with the conservation of historic landscapes. One of the problems we currently face is that we do not yet treat the historic environment as part of the natural environment, and we do not, therefore, integrate it completely with broader planning aims.

CONCLUSIONS

So where do we go from here? Within the historic environment sector, our mission over the last 50 years has been to achieve the conservation of the best of the past — what we value most — for the people of the future. Today, the watchword of the historic environment sector is 'value': not simply to protect the past for its own sake, but to realise the relevance and real value of the historic environment to the people of today and tomorrow. Once one accepts that the historic environment is relevant and valuable, then it must be managed.

I will highlight three areas of research which the historic environment sector is focusing upon at present.

The first relates to the scientific basis behind decisions about which sites, landscapes or monuments to protect. The sector still needs to learn from basic research into the interactions between structures and soils and climates, in order to ensure that decisions to preserve a particular historic building or a particular buried archaeological site are themselves sustainable. Since buried archaeological remains are subjected to change, there is no such thing as a stable burial environment. Nonetheless, it is seldom possible to model the behaviour of ancient remains over long periods of time. There is a need for continuing research since planning archaeologists and local authorities are required to make decisions every day about 'preservation *in situ*'. Yet they do not have the assurance that deposits which have been 'preserved', say beneath a new development, will survive without further degradation for the next 50 years, let alone the next 5000 years.

The inherent difficulty in decisions about preserving archaeological and historic remains is thrown into even sharper relief by problems associated with climatic change. Ancient buildings, parks, gardens and archaeological sites are threatened by the same dangers as the contemporary natural environment: flooding, coastal erosion, storm damage. But they are also vulnerable to other threats: historic buildings are extremely permeable; changes in moisture content occur rapidly and activate devastating cycles of salt crystallisation. The Centre for Sustainable Heritage (Bartlett School of Graduate Studies) has recently been commissioned by English Heritage to carry out a pilot study to develop a method for understanding and assessing climate change risk, and comparing local susceptibility with patterns of projected local climate change (Cassar, Pender *et al.* 2003).

The second direction for future effort concerns the theoretical approaches to selecting remains or sites for protection. A value-led approach to conservation is clearly needed, where the relevance of the historic environment is not merely recognised but the socio-economic values identified and extracted and 'returned' to the people to whom it belongs.

The third direction relates to the need for heritage-management decisions to be linked, or 'joined-up' with other considerations on the natural environment including spatial development, noise, waste, air quality, energy, biodiversity and so on. As new environmental strategies are established for London, it is imperative that they are consistent with one another and embrace the historic environment as part of the natural environment so if it is to be the exemplary, sustainable world city and leader in urban regeneration that the draft London Plan describes.

With the launch in March 2003 of the Mayor's Culture Strategy for London, this need is made even clearer. The Culture Strategy focuses very much on London's cultural diversity, examines the flavours of different boroughs, and states that '*A strategy for culture will only succeed if it recognises the connections and benefits culture can bring to a wider agenda ... as part of the broader infrastructure that sustains London. That means placing culture in the context of economic policy, plans for land use and transport development.*' We may go on from this to suggest that the next step must be to integrate cultural strategy with environmental strategy.

History is not only full of lessons for us but it is vital to our growth and future. We will be negligent if we do not manage our unique and tremendously rich historic legacy hand-in-hand with the natural environment as an *asset*, not a *constraint*. We must harness the power of the past to enrich the spirit of the future.

REFERENCES

Beier, A. L. and Finlay, R. (eds.) (1986) *London 1500–1700: the Making of a Metropolis*, London.

Bradley, K. R. (1994) *Slavery and Society at Rome*, Cambridge.

Cassar, M. and Pender, R. with Bordass, B., Corcoran, J., Hunt, Lord J., Nixon, T., Oreszczyn, T. and Steadman, P. (2003) *Climate Change and the Historic Environment, Scoping Study Report for English Heritage Commissions* (PNUM 3167), unpublished report, April 2003.

Clark, K. (2004) (*forthcoming*) Between a rock and a hard decision: the role of archaeology in the conservation and planning process. In *Preserving Archaeological Remains in situ?* (Nixon *et al.* eds.), MoLAS Monograph 2004.

English Heritage (2002) *State of the Historic Environment*, 91pp.

European Council (2000) Archaeology and the Urban Project — a European code of good practice (9 March 2000) (Cultural Heritage Committee: Activity, Archaeological heritage in urban development policies CC-PAT (99) 18 rev 3).

Evans, J. G. (1975) *The environment of early man in the British Isles*, London.

Finlay, R. and Shearer, B. (1986) *Population Grown and Suburban Expansion* in 'London 1500–1700: the making of a metropolis' (eds. A. L. Beier and R. Finlay), 37–59, London.

Graham, B., Ashworth, G. J. and Tunbridge, J. E. (2000) *A Geography of Heritage — Power, Culture and Economy*, 284pp Oxford University Press, New York.

Greater London Assembly (2003) *Cultural Captital: Realising the Potential of a World Class City*, Greater London Authority, June 2003.

Hulme, M., Jenkins, G. J., Lu, X., Turnpenny, J. R., Mirchell, T. D., Jones, R. G., Lowe, J., Murphy, J. M., Hassell, D., Boorman, P., McDonald, R. and Hill, S. (2002) *Climate Change Scenarios for the United Kingdom*, The UKCIP02 Scientific Report, Tyndall Centre for Climate Change Research, School of Environmental Sciences, University of East Anglis, Norwich, UK, 120pp.

HM Government (2002) *Achieving a Better Quality of Life, Review of Progress Towards Sustainable Development — Government Annual Report 2002*, DEFRA Publications, London. http://www.sustainable-, development.gov.uk/indicators/headline

IUCP, The World Conservation Union (1992) The Earth Summit, United Nations Conference on Environment and Development (UNCED).

Mayor of London (2002) *The London Plan, Spatial Development Strategy*, Consultation draft, June 2002.

Millett, M. (1990) *The Romanization of Britain*, an essay in archaeological interpretation, Cambridge University Press, 255pp.

National Trust (1999) *Valuing our Environment*, A study of the economic impact on conserved landscapes and of the National Trust in the South West 1998.

Nixon, T., McAdam, E., Tomber, R. and Swain, H. (eds.) (2003) *A Research Framework for London Archaeology 2002*, Museum of London 2002.

Sidell, J., Cotton, J., Rayner, L. and Wheeler, L. (2002) *The Prehistory and Topography of Southwark and Lambeth*, MoLAS Monography 14, Museum of London Archaeology Service, London, 98pp.

Woolf, G. (1998) *Becoming Roman, the Origins of Provincial Civilisation in Gaul*, Cambridge.

Environmental Developments
and Perceptions

Chapter 5

Environmental Strategies for London

David Goode

INTRODUCTION

The Mayor of London is required by the Greater London Authority (GLA) Act 1999 to produce a series of strategies that together provide the basis for London's future development. These eight strategies include the overarching London Plan that sets the scene for the next 20–30 years, together with detailed policies for transport and economic development. Alongside these are four environmental strategies; biodiversity, municipal waste management, air quality and ambient noise. The Mayor is also required by the Act to produce a culture strategy. Shortly after his election Ken Livingstone also agreed to produce an energy strategy as he recognised that this is an essential element of sustainable development.

The main elements of each environmental strategy are reflected in the overall London Plan and, where appropriate, in the Transport and Economic Development Strategies. This series of plans provide the basis for improving London's environment and also provide an integrated framework for sustainable development. Thus, whilst improvement of London's immediate environment, by reducing pollution and improving the quality of life for Londoners, is the main purpose of the environmental strategies, this is not the sole objective. The strategies also address London's wider impacts on the global environment and identify action to reduce damaging or unsustainable processes.

The development of each strategy has included widespread and lengthy public consultation and all the strategies have now been adopted and published. One of the requirements of the GLA Act is that the Mayor should endeavour to ensure that all his strategies are consistent with one another. This ensures that each forms part of a holistic approach covering all the strategies. This is a daunting task, and one that has probably never been achieved previously by any national or city government. Since the London Plan provides the framework and statutory basis for implementation of policies, particular effort was made to ensure that all the key environmental policies are contained in the London Plan and that there is consistency across all the strategies.

SUSTAINABLE DEVELOPMENT

The general functions of the Greater London Authority, as defined in the GLA Act, provide the basis for the three main components underpinning sustainable development in London. That is, economic development, social development and promoting improvement of the environment in London. However the legislation goes much further by placing a duty on the Authority to consider the implications for sustainable development, notably in its strategies, but also in any decisions that are made. Recognising the disproportionate influence that London has on the economy and environment outside its boundaries, the Act mandated the GLA to exercise its powers in ways which it considers best calculated to contribute to sustainable development in the United Kingdom. To meet these obligations makes it even more essential for the Authority to develop consistent policies for sustainability.

Before examining the way in which the Mayor's strategies address sustainable development it is relevant to consider two important concepts that have helped us to understand the magnitude of the environmental problems involved and to identify appropriate sustainable solutions for the future. These are the ecological footprint and city metabolism.

William Rees (1992) coined the phrase 'ecological footprint' in his study of the Vancouver region. He demonstrated that the level of resource use of that part of Canada, expressed on a per capita basis, would require the resources of two additional Earths if extrapolated to the entire world population of 6 billion people. Rees not only demonstrated the gross inequity of resource use on a global basis but also showed that the ecological footprint of a major western city is spread across the whole world.

Since that time, numerous footprint studies have been undertaken. A study published in 2002 estimated London's ecological footprint to be 293 times the size of the city itself, or about the size of Spain (City Limits Report). The per capita footprint in terms of useable global resources is approximately 6.3 global hectares, whereas our earth share is only 2.18 global hectares. These figures have been broadly substantiated by a more recent report commissioned by London First on behalf of the business community, which showed a per capita footprint of 5.8. This means that London's population is considerably more profligate in its use of resources than its fair share would allow on a global basis. We all know this in general terms but these detailed footprint studies help to quantify the scale of the problem. Effectively we need three Earths to support this level of resource use. As Mayor of London, Ken Livingstone has frequently quoted these figures and he recognises the need to reduce our global footprint if London is to become truly sustainable.

The second concept, which I find particularly helpful for identifying practical strategies, is to view the city as an ecological system and to

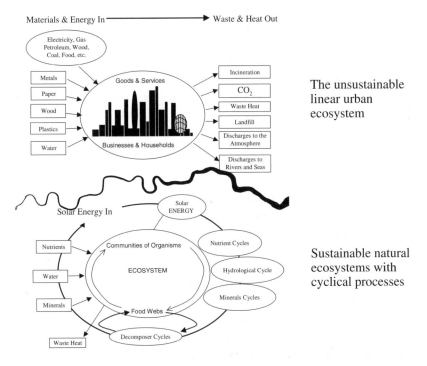

The unsustainable linear urban ecosystem

Sustainable natural ecosystems with cyclical processes

Fig. 1 The Ecosystem analogy for London.

consider its various functions as part of an 'urban metabolism'. For those familiar with ecological systems it is all too obvious that most urban systems are inherently unsustainable because this urban metabolism is linear rather than cyclical. By contrast many natural ecosystems have a series of inbuilt circular processes, preventing most wastage which enables them to reach a stable and sustainable equilibrium. This almost entirely one-way process is particularly marked for affluent cities in developed countries whose 'metabolisms' consume vast quantities of material imported daily for personal and industrial use, and discharge waste products as unwanted residues. A great variety of materials are needed such as paper, wood, plastics, water, fuel aggregates and also energy. After these are fed into the system, waste in the form of solids, liquids and gases emanate in many different forms at the end of the process. A particular feature of London, which is no exception to this picture, is that most of the solid waste is dumped as landfill way outside the city. Carbon dioxide emissions to the atmosphere resulting from the power requirements of buildings and from energy used by transport have a number of environmental consequences. These contribute to global warming and sea level rise. More immediately, nitrogen dioxide and fine particles emitted by traffic reduce air quality and cause respiratory problems, especially for the elderly and those suffering from asthma. There are also significant discharges of liquid pollutants to rivers and streams which immediately affect biodiversity of the urban environment. The hard surfaces of the urban fabric themselves exacerbate the problem of flooding owing to accelerated run-off. The challenges for London in reducing these sources of pollution and adverse environmental impact are discussed in more detail below and are also reviewed in the Mayor's *State of the Environment Report for London* (published in May 2003).

In seeking to find sustainable solutions for towns and cities, how can the knowledge and experience of natural systems be brought to bear on the management of urban areas? Over half the world's population now live in urban areas and it is predicted that by 2030 this will rise to about two thirds, with the correspondingly greater use of resources and production of waste that characterises our current 'urban lifestyle'. Ken Livingstone is on record as stating that we cannot achieve sustainable development without sustainable cities. The idea of a compact sustainable city is at the heart of his London Plan which will now provide the strategic planning framework for London for the foreseeable future. It is argued that a more compact city offers greater opportunities for sustainable solutions in terms of transport, energy use, and the whole metabolic process. As I have argued elsewhere (Goode 2000) cities, rather than being perceived as the problem, can provide the solution to sustainable development. There is no doubt that it is in cities that the greatest leverage exists to make effective changes towards sustainability. Compared with rural areas, towns and cities can be far more efficient in their use of resources such as energy because there are enormous savings to be made through economies of scale resulting from their high densities. Cities also offer considerable opportunities to put solutions into practice and explore new approaches through the involvement of communities and local action, as stimulated by the international movement initiated by the Earth Summit at Rio in 1992, known as 'Agenda 21'.

Benefits of the compact city were well illustrated by Newman's (1996) classic study of 32 major cities in North America, Europe, Australia and Asia in 1996. He showed that cities could be divided into distinct categories related to their density and overall level of fuel consumption for transport. This was effectively a measure of dependence on the private car as opposed to public transport. At one end of the spectrum are extremely low-density cities such as Houston, Texas and Phoenix, Arizona where fuel consumption per person is five times that of London. Most European cities have moderate densities but low levels of fuel consumption, whilst the Asian cities examined, such as Tokyo and Singapore, have a significantly greater density and lower levels of fuel consumption. This study demonstrated that economically successful and desirable cities can have low levels of fuel consumption, reflecting less car dependence; for example Vienna, Copenhagen and Stockholm which use only 10 to 20% of the fuel required by American cities.

Another form of energy economy, that Scandinavian cities have perfected, is by the use of community heating systems. A recent analysis of the potential for such schemes in the UK demonstrated that London offered 27% of the UK capacity. Again this is related to the economies of scale stemming from higher urban density.

These are just two ways in which a compact city can be an advantage in terms of sustainability. They illustrate the point that viewing the city in terms of its metabolism not only helps to understand how it functions but also helps to identify and prioritise practical measures to improve our environmental performance, and thereby reduce our global ecological footprint. This is crucial if we are to be successful in combating climate change and reducing London's global impacts on biodiversity and natural resources.

THE MAYOR'S STRATEGIES

It is clear, therefore, that the Mayor's Environmental Strategies have a dual role. Whilst improvement of London's immediate environment, by reducing pollution and improving the quality of life for Londoners, is their main purpose, this is not their sole objective. Evidence from ecological footprints studies and city metabolism demonstrate ways in which a city such as London can offer real solutions to long-term sustainability. So how do the Mayor's environmental strategies contribute to such a vision?

The Mayor's London Plan makes it clear that to become an, "exemplary, sustainable world city", London must use natural resources more efficiently, increase its reuse of resources and reduce levels of waste and environmental degradation. As London grows, these objectives will become ever more important. The shift towards a compact city, which is inherent in the London Plan, will contribute towards these objectives. It will enable more efficient use of resources such as land and energy and will also enable the 'proximity principle' to be applied to promote greater self-sufficiency. Implementing the Mayor's environmental policies will enable London to become a more sustainable and self-sufficient city, healthier to live in and more efficient in its use of resources. It should also be a better neighbour to its surrounding regions by consuming more of its own waste and producing less pollution.

AN ENERGY STRATEGY FOR LONDON

As with any natural system, energy is a key element of London's metabolism. Its supply and use have to change as part of the strategy of achieving long-term sustainability. London's impact on climate change and global temperatures, with the associated rise in sea level, is central to the issue and has determined much of the policy content of the energy strategy. If London is to make a significant contribution to the reduction of greenhouse gas emissions we need to substantially restrain our energy use and promote the use of renewable energy. However, the Mayor's Energy Strategy goes further than this. It aims to ensure that energy is produced and used in an equitable and sustainable way. Achieving this will not only minimise London's contribution to climate change but also help to give everyone access to affordable energy services and, most significantly, promote the development of our green economy.

As a city, London consumes more energy than Ireland and about the same as Greece or Portugal. London's population is now growing faster than the UK as a whole. This is driving increases in energy consumption, in domestic buildings, offices and the transport system. All of these outstrip the national rate of growth in energy demand. Projections in the London Plan suggest that London's population of 7.3 million will increase by a further 800 000 people, equivalent to a city the size of Leeds, by 2016. This projected growth poses challenges, but also offers enormous opportunities for developing new solutions in the energy field to meet both environmental and social objectives.

On climate change there is a broad international consensus that human activity is altering the global climate through emissions of greenhouse gases, with potentially serious consequences for humankind worldwide. Since 1992 international efforts have been made to secure agreements to stabilise and reduce greenhouse gas emissions. The Intergovernmental Panel on Climate Change (2001) concluded that emission reductions far in excess of those agreed at Kyoto will be necessary during the 21st Century. The Royal Commission on Environmental Pollution recommends a target for reducing carbon dioxide emissions in the UK by 60% relative to 2000 by 2050. The Government accepted these finding in its Energy White Paper of February 2003.

Even with these efforts to reduce emissions, some impacts of climate change are now inevitable. In October 2002 the London Climate Change Partnership published a report identifying key ways in which London is likely to be affected by climate changes. These impacts include significantly increased summer temperatures and greater frequency of intense storms. It is expected that this will lead to increasing demand for electricity and water and a decrease in the comfort and safety of buildings and transport infrastructure. The study highlighted the increased risk of flooding, both in terms of heightened probability and the huge economic impact this could have due to the value of assets located in areas of flood risk within the capital. These are risks that are already with us, and there is no doubt that action will be needed to alleviate these problems. However, London also needs to take a lead in developing alternative forms of energy in the longer term to minimise our contribution to climate change.

But this is not the only reason why the Mayor has produced an energy strategy. There is also the problem of fuel poverty. A significant number of people in London have to spend an excessive proportion of their income on energy supplies for heating their home. As a result many households are unable to maintain healthy indoor temperatures in winter. Households in this situation are defined as fuel poor and it is estimated that this applies to 4–500 000 households in London.

Fuel poverty in winter is caused by a combination of low incomes, poorly insulated and often under-occupied housing, inefficient heating equipment, and energy pricing and payment structures that tend to penalise customers who use less energy. Living with temperatures below the recommended minimum can damage health and sometimes results in death. These risks are greater for people on lower incomes, children, older people and people with disabilities. Fuel poverty also affects the wider community through increased health expenditure and damage to local economies.

As global temperatures rise there will be an increasing number of occasions with excess summer temperatures in inner city areas, which will seriously affect the health of the elderly and sick. Such episodes already cause thousands of deaths in cities in the US and China, and may occur in London by mid century. This will have to be considered in planning and housing, as well as energy policy. The Mayor's energy strategy provides a vision for London's energy use in 2050 as a goal to work towards which is in line with the Royal Commission's recommendations and which will also address the key issue of fuel poverty. Achieving this vision provides a huge opportunity for delivering economic, social and environmental improvements. In particular the strategy aims to:

- Minimise the impact of London's energy production and use on people's health and on the local and global environment,
- Reduce London's contribution to climate change by minimising emissions of carbon dioxide from all sectors (commercial, domestic, industrial and transport) through energy efficiency, combined heat and power, renewable energy and development of the hydrogen economy,
- Help eradicate fuel poverty, giving all Londoners (particularly the most vulnerable groups) access to affordable warmth. In future excess temperatures will have to be considered,
- Contribute to London's economy by increasing job opportunities, skills and innovation in delivering sustainable energy and improving London's housing and other building stock.

The strategy proposes that London should reduce its emissions of carbon dioxide by 20% relative to the 1990 level by 2010 as the crucial first step on a long-term path to a 60% reduction from the 2000 level by 2050. This will ensure that London plays its full part in meeting national targets for cutting carbon dioxide emissions. The GLA has produced the first detailed inventory of London's energy use and greenhouse gas emissions. Figures were published in the Mayor's State of Environment Report in 2003 and will be updated regularly.

Innovative housing developments such as Bedzed, in South London at Sutton, show that it is quite practical to integrate energy efficient design of buildings with renewable energy production, so that its operation results in no carbon dioxide emissions to the atmosphere either from the building or from power stations providing energy. Local surveys have attested that buildings and surrounding landscape form an attractive living environment for the occupants and the neighbourhood. The strategy proposes that there should be at least one zero-carbon development in every London borough by 2010. To achieve this each borough will need to identify at least one suitable site for such a development and include these sites in their Unitary Development Plans. Developments of this kind will provide examples of good practice which can be emulated elsewhere.

The strategy aims to develop new technologies and encourage the use of renewable energy as a major part of London's energy programme. A huge opportunity exists for London to deploy renewable energy across the capital and to purchase green power generated outside the capital. In the long-term the Mayor wants renewable energy technology to make a major contribution to London's economy and energy supply mix. The strategy includes detailed medium-term targets to meet these objectives.

Major planning applications referred to the Mayor are now required to incorporate a range of features to reduce dependence on fossil fuels. These include passive solar design, natural ventilation, borehole cooling and use of vegetation on and around buildings. An assessment of the energy demand of proposed major developments should become the norm along with the incorporation of appropriate renewable energy technology.

One way in which London could improve its environmental performance and at the same time improve people's quality of life is through the development of combined heat and power schemes. Increased use of combined heat and power (CHP) is an effective way to reduce carbon dioxide emissions and could also contribute significantly to eradicating fuel poverty. The heat generated from CHP plants can be used in industrial processes, or for heating or cooling buildings via community heating networks. These can provide affordable warmth to large numbers of homes, helping to tackle fuel poverty on a significant scale. It is proposed that the amount of CHP capacity available in 2000 should be doubled by the year 2010. The Borough of Woking just outside London has pioneered an integrated approach involving conventional CHP, renewable energy production (including use of fuel-cell technology and photovoltaics), and its own local grid (or private wire network) to optimise costs. Will such local networks play an increasing role in the future?

MUNICIPAL WASTE MANAGEMENT

Waste is another area where we need to significantly improve our efficiency. It is not simply a matter of improving levels of recycling, which is how the problem is often perceived. If London is to become sustainable, a more fundamental long-term change is required to establish a secondary materials economy. We need to develop a new business culture, where components of the waste stream are automatically considered as potential products for new industries. The policies contained in the Mayor's strategy, published in September 2003, set the framework for such a change. Substantial progress has already been made through the London Remade Programme, funded by the London Development Agency, and this approach is now being promoted more widely as a component of economic development. The Mayor's Green Procurement Code is another key initiative which provides the necessary link between environmental improvement and business performance.

The present situation in London is that we are producing about 17 million tonnes of waste each year. Of this about 25% is municipal waste. A relatively small amount (approximately 8%), of municipal waste is currently recycled and London is lagging way behind many other European countries in this area. About 19% of municipal waste goes for incineration, from which 64 MW of energy is generated in two main incinerators, one in north London at Edmonton and one just east of London Bridge in Lewisham. However, most municipal waste (73%) goes to landfill, mostly in counties outside London. Some goes by barge down the Thames to landfill sites in Essex, but most goes by train further afield. This unsustainable practice is unacceptable to the surrounding counties and to central government. Following investment from government of £50 million over recent years most boroughs are now expanding their facilities and services for recycling, in some cases quite rapidly. However the continued growth, at about 2% per year, in the total amount of waste that they are dealing with, adds considerably to the challenge.

London has to work within the European Union (EU) Directive on landfill that requires that alternative means of dealing with waste are installed by 2010. This change is necessary anyway because landfill capacity in the surrounding regions is falling dramatically. For both these reasons London needs to shift towards more sustainable methods for managing its waste. This begins with each London borough being required to achieve statutory targets for recycling household waste, averaging 25% across London by 2005–6. Some are already achieving these targets, but others will find it a challenge, especially inner London boroughs with a high proportion of multi-occupancy buildings and a more transient population. Another challenge is the urgent need to find politically acceptable and technically feasible alternatives to the controversial practices of incineration and landfill to deal with both residual and hazardous waste.

The Waste Strategy has identified practical programmes for tackling these significant problems over the next ten years. Firstly it is planned to establish waste reduction and reuse programmes working with retailers, manufacturers and London Boroughs with the objective of gradually reducing the total amounts of waste being produced. In order to promote higher levels of recycling the strategy aims to ensure that every household in every Borough will have a kerbside collection for at least three materials by September 2004. This, as current practice has shown, will substantially increase the amount of recyclable material that is collected. There will also be a revolution in the use of Civic Amenity Sites so that they become re-use and recycling centres, with the emphasis on re-use of waste rather than waste disposal.

One of the most important elements in the Mayor's strategy is to develop new recycling industries. The Mayor has emphasised the importance of converting the waste stream into new industries, thereby creating new jobs. This will be a major responsibility of the London Development Agency (LDA), one of whose key objectives is to develop the green economy. Over the last three years, the LDA has invested £5.6 million into the development of new industries and markets, by promoting *London Remade*, with particular emphasis on paper, glass and organic materials. This has already been remarkably successful in diverting 268 000 tonnes from landfill and creating another 150 000 tonnes of reprocessing capacity. The scheme has also involved some 500 people receiving training in new skills. The GLA is setting up a Markets Taskforce to promote new opportunities of this kind.

Other materials that could be dealt with in this way include the kind of plastics that everyone at times puts into a bin somewhere. Currently most of that is not going to be recycled, but will be going into landfill or some other form of disposal. However through recycling plastic there is a significant business opportunity for London, either to produce pellets or to produce plastic bottles. Other new emerging technologies include Glasphalt, a mixture of rubber from tyres and glass. In this case glass is used as an aggregate, turning it into granules and it is then mixed with rubber to produce quiet road surfaces, thus resulting in the dual advantage of reducing waste and reducing environment noise.

Other initiatives for waste and energy have emerged from business and other branches of the public sector, partly in response to the Mayor's Green Procurement Code for all London businesses. All the London Boroughs and about 250 businesses and other organisations have signed up to this code.

So these are some example of ways in which the Municipal Waste Strategy will encourage development of the cyclical processes referred to earlier.

BIODIVERSITY

Conservation of biodiversity, which is addressed in detail in the Mayor's Biodiversity Strategy and in the London Plan, is one of the essential elements of his environmental strategy. The subtitle of the strategy, *Connecting with London's Nature*, emphasises the social context, since one of the main objectives of the strategy is to ensure the conservation of London's natural heritage for people to enjoy. This is also important for their health and well-being, and generally for their environmental understanding and awareness (see also Chapter 8). The Mayor has adopted the well-established procedures for identification and protection of important habitats in London as the basis for his Biodiversity Strategy, which was published in July 2002. At present, London is the only part of Britain where there is a statutory requirement for a biodiversity strategy as part of regional planning and it may provide a useful model for other towns and cities in the UK. The strategy also has an international dimension by making proposals to clamp down on the illegal international trade in endangered species for which London's airports are one of the main points of entry to Europe.

Policies for protection of habitats and species through strategic planning are largely based on the voluntary procedures adopted by the London Ecology Committee over the period from 1986 to 2000. This involves identification of a hierarchy of protected sites from those of London-wide importance to others of borough or local importance. The Mayor has now identified over 130 Sites of Metropolitan Importance for Nature Conservation which together provide the London-wide strategic framework for the protection of habitats in the capital. The London Plan states that these should be given strong protection in Unitary Development Plans. The Mayor also expects boroughs to identify Sites of Borough and Local Importance using the criteria and procedures set out in the Biodiversity Strategy. This has now resulted in over 1500 sites being identified for protection in Local Plans, representing almost 20% of London's land area. Whilst this has been done previously on a voluntary basis the requirements of the London Plan now bring this area of work firmly into the statutory planning process.

The Biodiversity Strategy also deals with a range of other topics which can be summarised as follows:

- management, enhancement and creation of habitats,
- greening within urban regeneration (including green roofs and effects on urban climate),
- improved access to nature for Londoners,
- education and awareness raising,
- promoting the green economy, including the importance of tourism, through the London Development Agency,
- London's international role in biodiversity conservation,
- identifying London's biodiversity footprint.

The whole Strategy depends upon partnerships for its implementation, in particular the London Biodiversity Partnership which has developed a series of Action Plans for priority habitats and species. The Mayor's sees this partnership as a crucial mechanism for implementation of his strategy through the wide range of stakeholders involved. The long-term success of the Biodiversity Strategy will be measured against two targets, to ensure:

- That there is no net loss of important wildlife habitat,
- That a net reduction is achieved in the Area of Deficiency of accessible wildlife sites.

The Biodiversity Strategy addresses key issues which are fundamental to the quality of life for Londoners. Our success in conserving and monitoring biodiversity can also be used as a valuable indicator to measure London's achievements with regard to sustainable development. The London biodiversity database is probably the most detailed of any region of the UK and will provide the basis not only for detecting and assessing trends but also for predictive modelling. This will be particularly important as changes occur to London's environment and local climate in response to London's overall growth, or to global climate change.

AIR QUALITY

Clearly one of London's main environmental problems is air quality. Although London no longer suffers the smogs of the 1950s, the quality of the air is below the standards expected in terms of health and environmental quality. The main problems are emissions from road traffic in the form of nitrogen oxides and air-borne particles (see Chapter 7). London currently fails to meet EU and national targets for air quality because of the size of the conurbation and because of the large amount of road traffic. The Mayor's Air Quality Strategy makes detailed proposals to meet the legal targets, and to promote longer-term solutions by introducing cleaner vehicle technologies. Actions being taken include the following:

- Continuing to improve the Transport for London (TfL) regulated bus fleet so that it is the cleanest in the UK. The Mayor has set emissions standards to be met in 2005,
- Piloting 3 'zero emission' hydrogen fuel-cell powered buses in January 2004 and 10 diesel-electric hybrid buses in September 2004,
- Reducing taxi emissions through imposing licensing conditions (to be published in 2004),
- Reducing traffic growth by improving the provision of public transport, and encouraging developers to make easy access to public transport a normal part of new developments. Introduction of the congestion charge in central London in February 2003 reduced traffic by over 20%,
- Investigating the feasibility of low emission zones — prohibiting the most polluting vehicles from specified areas,
- Publishing guidance for fleet operators and encouraging businesses to reduce exhaust emissions from their vehicle fleets,
- Working with London boroughs to help them reduce air pollution,
- Settings standards to improve energy efficiency in new homes and offices and thereby reduce indoor emissions.

As with the implementation of more sustainable energy policies, air quality improvement will also depend on the progress made by the London Development Agency in promoting new technologies producing less pollution, such as hybrid gasoline-electric engines and hydrogen fuel cells. There are other incentives too. Vehicles using such power sources are exempt from the congestion charge. There is, of course, a strong link here between the strategies to achieve long-term solutions which meet several objectives, improving quality of life for Londoners and reducing our dependence on fossil fuels.

AMBIENT NOISE

Strategic policies and practical measures to deal with noise in urban areas have until recently had lower priority than dealing with other environmental problems. Action has been less advanced in the UK than in some other European countries. However, the requirement for the Mayor to produce the UK's first city-wide strategy for tackling environmental noise has resulted in much progress over the past three years. The main focus of the Mayor's London Ambient Noise Strategy (2004) is on reducing noise through better management of transport systems, better town planning and better design of buildings. This includes minimising noise on roads and railways, and considering where noisy activities are best sited to ensure that we protect housing, schools, waterways and open spaces.

Ambient or environmental noise is long-term noise from transport and industry, as distinct from noise caused by neighbours, construction sites, other local nuisances, or noisy workplaces. Local 'nuisance noise' from noisy neighbours, pubs or clubs, road works, or construction sites, is dealt with by local boroughs. This is not part of the ambient noise strategy.

The strategy sets out the main steps that need to be taken, including quieter road surfaces, smoother traffic flow, rail infrastructure improvements, aircraft noise measures, and investment in improved design for new developments. This will be supported by new data currently being produced by the London-wide noise-mapping scheme funded by central government. (An example of such computer modelling was displayed at the LEAF conference by Stocker (2003).)

The European Environmental Noise Directive, 2002/49/EC, places new duties on the UK government, including noise mapping. Noise maps will show the location of noisy and quiet areas. Maps will show at local level the numbers and locations of people exposed to different noise sources, including roads and railways. Following work on noise mapping, on the effects of noise, and on the cost-effectiveness of various ways of reducing noise, the UK government aims to produce a national ambient noise strategy by 2007.

London was chosen by central government as the first area for this initiative in the UK and work commenced in 2003. Road traffic noise-mapping will be completed during 2004 for the whole of London. The Mayor has also launched a survey to measure diurnal noise levels in selected areas of inner London. Some attitude-survey data is also available from the London Household Survey 2002. All this new information will greatly assist the Mayor in implementing his ambient noise strategy. The strategy will lead the way in developing new ways of dealing with noise in cities, at a time when international pressure is growing to take more action.

KEYS TO FUTURE SUCCESS

Successful implementation of the policies contained in these environmental strategies will depend on collaboration between GLA and a wide range of stakeholders responsible for all the different aspects of environmental management in London. The following are the key elements:

* Working through the London Development Agency to develop skills and economic measures for sustainable solutions such as new green technologies,
* Integrating environmental polices across all aspects of urban regeneration,
* Finding ways to embed and promote locally developed solutions in the wider framework of strategic planning and regional development,
* Working with Transport for London to mainstream environmental improvements in London's transport systems,
* Building effective partnerships for implementation of environmental policies with stakeholders across the voluntary and private sectors,
* Developing joint approaches between the GLA, business, non-governmental organisations (NGOs) and London's academic community to investigate existing barriers to progress, and to promote research into innovative solutions,
* Promoting opportunities for changes in urban design and lifestyles to encourage sustainability, such as zero-energy housing developments,
* Incorporating biodiversity within the urban fabric as an essential ingredient of quality of life in a more compact city.

The overall effect of the Mayor's five environmental strategies over the next twenty years will be to make significant improvements in our own local environment as well as reducing London's wider global impacts. The strategies provide many of the essential ingredients to make London a truly sustainable world city.

REFERENCES

Anon (1999) *Greater London Authority Act 1999*, The Stationary Office.
Anon (2002) *City Limits: A Resource Flow and Ecological Footprint Analysis of Greater London*, Best Foot Forward Ltd., Oxford.

DTI (2003) *Our Energy Future-Creating a Low Carbon Economy*, Department of Trade and Industry. The Stationary Office, London.

Goode, D. A. (2000) "Cities as the key to sustainability," In: *Where Next? Reflections on the Human Future*, D. Poore (ed.), Royal Botanic Gardens Kew, London.

Intergovernmental Panel on Climate Change (2001) Third report, Cambridge University Press.

London Climate Change Partnership (2002) *London's Warming: The Impacts of Climate Change on London*, Greater London Authority, London.

Mayor of London (2002) *Cleaning London's Air: The Mayor's Air Quality Strategy*, Greater London Authority, London.

Mayor of London (2002) *Connecting with London's Nature: The Mayor's Biodiversity Strategy*, Greater London Authority, London.

Mayor of London (2003) *Green Capital: The Mayor's State of the Environment Report for London*, Greater London Authority, London.

Mayor of London (2004) *The London Plan: Spatial Development Strategy for Greater London*, Greater London Authority, London.

Mayor of London (2004) *Sounder City: The Mayor's Ambient Noise Strategy*, Greater London Authority, London.

Mayor of London (2004) *Green Light to Clean Power: The Mayor's Energy Strategy*, Greater London Authority, London.

Newman, P. (1996) "Transport: reducing automobile dependence," *Environment and Urbanisation* **8**(1), 67–92.

Rees, W. E. (1992) "Ecological footprints and appropriated carrying capacity: what urban economics leaves out," *Environment and Urbanisation* **4**(2), 121–130.

Royal Commission on Environmental Pollution (2000) *Energy: The Changing Climate*, 22nd Report of the Royal Commission on Environmental Pollution, The Stationary Office, London.

Stocker, J. & Carruthers, D.J. (2003) How accurate is a noise map created using air quality source data? *Acoustics Bulletin 28*, no. 2.

Chapter 6

Places, People and Socio-Economic Differences in Health

Michael Marmot and Mai Stafford

INTRODUCTION

People who live in deprived areas have higher death rates, worse health, and are more likely to be the victims of crime than people living in better off areas. But is it the area or the people? To what extent can we attribute this worse health and lower levels of well-being to the area? Or is their worse health a characteristic of the poor people who live in poor areas?

The account that follows of a fatal encounter between two young men illustrates the problem. It gives ample ground for focussing on the environment and the people within it. It starts with an everyday scene at a local authority housing estate in North London. The estate is an unlovely place to visit. One of those concrete architectural "masterpieces" that everyone loves to hate, it has a great deal of concrete and little green. Torn basketball nets, and a tatty room with a table tennis kit are not going to occupy the energies of thrusting young men, nor provide a belief in the future. There are a few boys standing in the foyer, smoking dope, urinating on the floor, spitting and tagging (writing graffiti) on the walls. There is nothing to do on the estate; little space physically and culturally. There are gangs but they appear rather low key. Two boys fought. Why? One had drawn a line through the other's graffiti. This was a grave insult. A boy's tag was his signature, and he had been demeaned. This time it ended differently. The insulted one went out with a kitchen knife. He confronted the one who had insulted him who now added injury to insult by knocking the aggrieved one to the ground. Out came the kitchen knife and he stabbed his antagonist twice through the heart. The boy died (Brockes 2001).

It is not difficult to fill in some of the background on the housing estate. Hardworking immigrant parents; long hours on low pay in menial jobs, trying to make a go of it in Britain. Half the households are on housing benefits. The children vary in the degree to which the system will serve them. Some will do better at school but, unlike their comfortable middle class neighbours, most will end up in low paying jobs, if in jobs at all.

The crime and violence of these young people have wider effects: older people's fear of going out and consequent isolation. Social isolation in turn increases the death rate of older people. The purpose of this chapter is to address the question of whether the characteristics of areas contribute to the health of people who live there, over and above the characteristics of the people. Conceptually, the question is whether the same individual, with the same degree of deprivation would have different levels of health if living in a different environment. In asking about the geographic distribution of health and disease can we distinguish the effect of the places in which people live, and work, from the characteristics of the people themselves? The answer is not easy, because although conceptually we can place a poor person in a different area, in practice the properties of an area are related to who lives there. There is not a clear distinction between the "environment" and "people". People are an important part of the environment.

THE LONDON ENVIRONMENT AND HEALTH

One hundred years ago, the relation of environment to health in a British city was plain. A combination of dirty water, unhygienic practices, and poor nutrition could account for high rates of infant mortality, for example. And they were high. In the English city of York at the start of the 20th Century, infant mortality was just under 250 per 1000 live births in the poorest areas. Even in the richest areas, among the servant-keeping class, it was 94 per 1000 (Rowntree 2001). One hundred years later, infant mortality in Britain was dramatically reduced. However, the social gradient in infant mortality persisted, ranging from 3.7 in Social Class I to 8.1 in Social Class V. In other words the richest people in 1901 had infant mortality an order of magnitude higher than the poorest people in 2000 (Marmot 2002). This should lead us to think about environment in a different way. It is highly likely that the high infant mortality of the servant keeping classes was related to poor standards of environmental hygiene. This has improved radically. It is unlikely that poor sanitary conditions account for much of the excess mortality and worse health of those at the bottom of the social hierarchy in 2000. The poorest have infant mortality rates of 8.1. This does not mean that environment ceases to be important. There will be features of the physical and biological environment that may still have an impact on health. More important, perhaps, will be the social environment.

In this paper, we review the geographic distribution of health with a focus on London, and then look at how the characteristics of areas and of the people who live in them may interact to cause ill-health. The paper ends by asking how we might characterise the social environment.

GEOGRAPHIC DISTRIBUTION OF HEALTH IN LONDON

Figures 1 and 2 show the life expectancy for men and women in London boroughs. The variation is quite marked. This led to the much quoted statistic: travel six stops east on the Jubilee line from the West End of London and life expectancy drops a year for each stop. As Figure 1 shows, men in Westminster and Kensington and Chelsea have life expectancy at birth of 77 years or more. Men in the East London boroughs of Tower Hamlets, Newham and Barking and Dagenham have life expectancy of less than 73. The trouble with a statistic like life expectancy is that it is difficult to grasp the import of five years of life expectancy. Abolishing coronary heart disease, the number one cause of death, and replacing it with nothing else, would add just under four years to life expectancy. A difference in life expectancy of five years for six stops on the underground train is, therefore, dramatic.

Among women, it is a similar pattern: lower life expectancy in the boroughs of East London, high in Westminster, Kensington and Chelsea, and Richmond upon Thames. The differences among women are less extreme than the differences among men.

Such dramatic differences are not unique to London. In Washington DC, in 1990, black men had life expectancy of 57.9. In Montgomery County Maryland, which is the suburban area adjacent to the city of Washington, life expectancy for white men was 76 (Murray *et al.* 1998). This highlights our central question. In comparing life expectancy in downtown Washington with that of suburban Maryland, we are comparing poor blacks with rich whites. If one judged that it was the property of poorness or blackness that

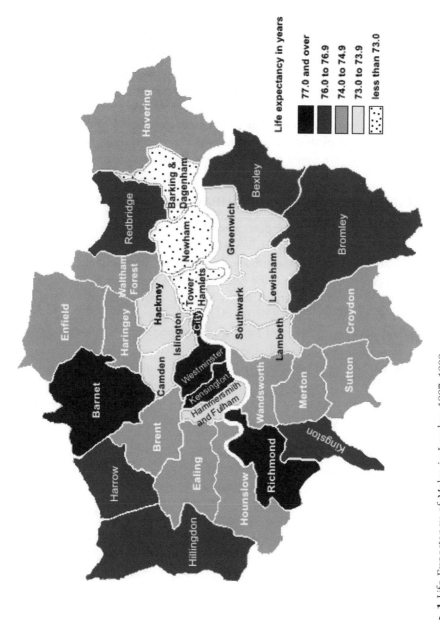

Fig. 1 Life Expectancy of Males in London 1997–1999.
Source: Office for National Statistics (Fitzpatrich & Jacobsen, 2001).

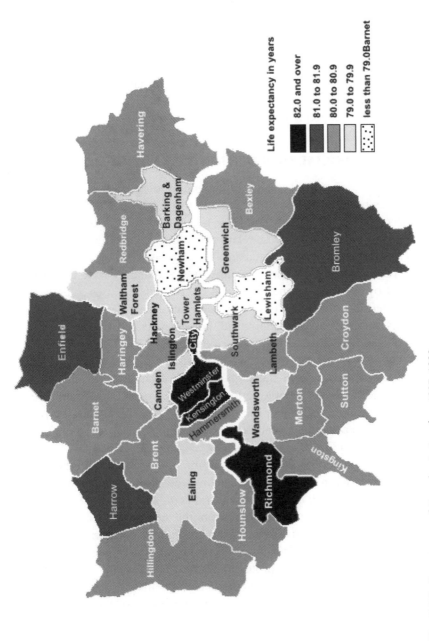

Fig. 2 Life Expectancy of Females in London 1997–1999.
Source: Office for National Statistics (Fitzpatrich & Jacobsen, 2001).

was somehow responsible for the life expectancy difference, then the geographic differences could be attributed to geographic clustering of individuals. Alternatively one might ask if one of the reasons that poor black people have worse life expectancy than rich whites is because of where they live.

Mortality is the end stage of a process. Morbidity, too, shows a geographical distribution that is remarkably similar to that of mortality rates. Poor health is more common in the East of London, less common in Westminster, Kensington and Chelsea.

HEALTH AND SOCIAL DEPRIVATION

A superficial knowledge of the social geography of London suggests that one obvious potential link with these patterns of health, is the pattern of social deprivation. The East End, for generations has been relatively deprived; the areas further west relatively, and absolutely, affluent. One way to look at the link more systematically is to classify areas according to their scores on a standard deprivation measure. The one we have used was developed by Peter Townsend (Townsend *et al.* 1988). It uses a few simple characteristics from the Census: the proportion in overcrowded living conditions, the percentage unemployed, home ownership, and household access to cars. Electoral wards are classified according to degree of deprivation. Figure 3 shows the results when wards are grouped into quintiles on the basis of deprivation: the higher the deprivation score, the more deprived. The percent of people reporting themselves in poor health rises in step wise fashion: the higher the deprivation score the greater the percent of people reporting themselves in poor health. These are national figures; the patterns in London are similar.

There is a body of work that now confirms this finding in other settings: the more deprived the area the worse the health (Diez-Roux *et al.* 1997, 2001; Davey Smith *et al.* 1998; Stafford *et al.* 2001). This does not, of course, deal with our introductory question: the area or the people? There is likely to be an interaction. This is shown clearly by a comparison of mortality rates by regions of England and countries of the United Kingdom separately for Social Class I (professional) and Social Class V (unskilled manual). The dramatic finding is the lack of variation geographically among those of the higher social class. In all regions of the UK people in Social Class 1 have a lower mortality rate than those in Social Class V, but the magnitude of the difference varies by region (Office of National

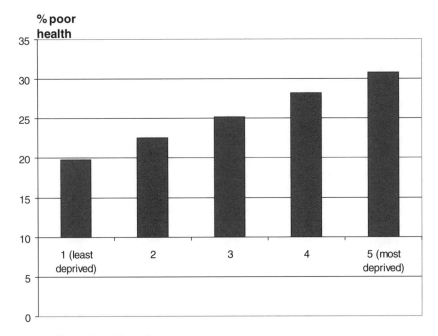

Fig. 3 Self-rated health and material deprivation.
Source: Data from Health Survey for England 1994–1999.

Statistics 2001). This is because there is a two-fold variation in mortality in Social Class V: highest in the north and Scotland, lower in the south. By contrast there is little regional variation in Social Class I. Putting it differently, the region in which they live matters much less, in terms of health, for a high status person than for one of lower status. A simple interpretation is that the environment matters more for people of low status. It may be that what it means to be low socio economic status (a combination of low educational attainment, low income, and associated type of occupation) varies from one region to another. This could relate to education, income, and occupation. These are characteristics of individuals, but educational opportunities including the quality of schools, the structure of incomes, and occupational opportunities are related to the social and economic environment.

ENVIRONMENT AND INDIVIDUAL SOCIO-ECONOMIC POSITION

We have, for many years, been investigating the causes of the social gradient in health: the fact that the lower someone's position in the social hierarchy, the worse is their health. Our studies of British civil servants, the Whitehall and Whitehall II Studies, show a clear gradient in health among people who are not materially deprived in the absolute sense of that word. In the Whitehall II Study, we have investigated whether knowledge of where somebody lives predicts markers of ill-health in addition to the individual's socio-economic position.

One could posit two alternative *a priori* positions. First, if you were of low status, living in a more deprived area would be a further insult to your health than if you were in a less deprived area. An alternative is that a poorer person, surrounded by others richer than he, would perceive himself to be lower in the hierarchy. His health would suffer as a result. The implication of the first is that health of a low status person would be worse in a more deprived area. The implication of the second is that his health would be worse in a less deprived area.

Table 1 shows a clear social gradient in poor health, poor mental health, and waist: hip ratio according to level in the occupational hierarchy. Waist: hip ratio is a predictor of coronary heart disease (Folsom *et al.* 1998). Areas were classified according to degree of deprivation on the Townsend score. Among low grade men and women, the more deprived the area, the worse the physical and mental health and the higher the waist: hip ratio. The relation with area is less clear among the high grades although, for the most part, this apparent interaction was not statistically significant. These findings are consistent with the first but not the second of the two *a priori* alternatives: people of lower status have worse health if they also live in a more deprived area. Living in the least deprived area appears to be better for their health but it does not raise it to the level of a high status person. It would appear that whatever area might be contributing to ill-health, there are characteristics associated with socio-economic position that are also important for health.

CHARACTERISING THE EFFECTS OF THE SOCIAL ENVIRONMENT ON HEALTH

As indicated, morbidity and mortality vary across different areas in London. For a given level of personal deprivation, residence in a more deprived area has additional health effects. Our work has shown that areas which are high in deprivation also tend to have a poorer quality social fabric or poorer social cohesion. For example, area deprivation is negatively correlated with social participation, contact with friends, integration into the wider

Table 1 Physical and mental health of participants in the Whitehall II Study by individual employment grade and deprivation in the area of residence, adjusted for age and sex.

	Townsend deprivation index		
	Least deprived (10th centile)	Middle (median)	Most deprived (90th centile)
% with poor self-rated health[a]			
high grades	8.8%	9.9%	12.3%
middle grades	12.3%	15.2%	22.2%
low grades	19.7%	24.4%	35.9%
% with poor mental health[b]			
high grades	10.1%	10.4%	10.9%
middle grades	12.9%	15.1%	20.1%
low grades	17.2%	20.3%	27.4%
Mean waist/hip ratio			
Men			
high grades	0.918	0.919	0.921
middle grades	0.921	0.926	0.935
low grades	0.930	0.937	0.950
Women			
high grades	0.786	0.787	0.789
middle grades	0.789	0.794	0.802
low grades	0.798	0.805	0.818

[a]Participants who reported that their health was average, poor or very poor were defined as having poor self-rated health.
[b]Participants who scored 5 or more on the 30-item General Health Questionnaire were defined as having poor mental health.

society, trust, attachment to the neighbourhood, tolerance and practical help (Stafford *et al.* 2003). We measured social cohesion in a selection of 95 electoral wards in London. A random sample of residents living in these wards received a postal questionnaire asking about their perception of the local social environment. Their responses were aggregated up to ward level and used to describe eight different aspects of social cohesion. Health data in these same areas were obtained from the Health Survey for England. The Health Survey for England is a series of annual surveys which are representative of the general population of England and are designed to monitor trends in population health and determinants of health. It includes about

15000 adults from all over England, but the analysis presented here is based on participants living in London only. Combining these two data sources gave us contextual information describing social cohesion at ward level and individual data on health and socioeconomic factors. Importantly, one set of respondents reported on their perception of the local social environment and a different set of respondents reported on their health status. This means that any associations seen here between social cohesion and health are not the result of people's tendency to report always positively or negatively.

Table 2 shows the association between aspects of social cohesion and health. For each of the eight social cohesion scales, areas were grouped according to whether they had low, medium or high scores based on tertiles of the distribution. Living in an area where residents are more trusting of each other was associated with a reduced risk of poor self-rated health. Compared with areas with the highest levels of trust, residents in low trust areas were 1.71 times as likely to rate their health as poor. This effect was over and above any relationship between individual characteristics (age, sex, social class, and economic activity) and self-rated health. Low levels of attachment were also associated with a greater prevalence of poor self-rated health.

Residence in an area where integration into wider society was lower was associated with an increased risk of poor mental health. Again, this effect was over and above individual characteristics.

These findings provide evidence that people living in areas with a poorer quality social environment are more likely to have poorer general health and poorer mental health (although we note that not all aspects of social cohesion are important). Over and above a resident's own socioeconomic position, living in an area where people are less trusting, feel less attachment to their area, and are less integrated into wider society carries risks for health.

CONCLUSIONS

In answer to the question posed earlier "is it the area or the people?", our work suggests that it is both. Living in an area with multiple social deprivation and with poor quality social cohesion increases a person's risk of poor health. Personal socioeconomic circumstances are also strongly associated with poor health. At particularly high risk are deprived individuals living in deprived areas. They experience the double jeopardy of personal deprivation combined with area deprivation.

Table 2 Health by tertile of social cohesion in London using data from the Health Survey for England 1994–1999 and from a new community questionnaire designed to capture the local social environment.

	Number of participants		Odds ratio (9% CI) of poor health compared with high social cohesion areas adjusted for age, sex and socioeconomic position	
			Lowest social cohesion	Medium social cohesion
Poor self-rated health[a]	1995	Trust	1.71 (1.18, 2.50)	1.55 (1.07, 2.24)
		Sense of attachment	1.79 (1.22, 2.63)	1.33 (0.91, 1.95)
		Tolerance/respect	1.41 (0.97, 2.06)	1.01 (0.68, 1.50)
		Practical help	1.29 (0.87, 1.93)	1.39 (0.94, 2.04)
		Contact with local family	0.92 (0.63, 1.36)	0.88 (0.60, 1.29)
		Contact with local friends	1.08 (0.74, 1.60)	0.95 (0.65, 1.39)
		Social participation	1.14 (0.78, 1.68)	1.03 (0.70, 1.51)
		Integration into wider society	1.41 (0.95, 2.08)	1.28 (0.88, 1.87)
Poor mental health[b]	1539	Trust	1.23 (0.86, 1.76)	0.84 (0.58, 1.22)
		Sense of attachment	1.11 (0.77, 1.59)	0.74 (0.50, 1.07)
		Tolerance/respect	1.16 (0.80, 1.68)	1.02 (0.07, 1.49)
		Practical help	0.89 (0.60, 1.31)	1.01 (0.70, 1.47)
		Contact with local family	1.27 (0.87, 1.85)	1.50 (1.04, 2.17)
		Contact with local friends	1.39 (0.96, 2.00)	1.15 (0.80, 1.67)
		Social participation	1.16 (0.80, 1.69)	1.02 (0.70, 1.48)
		Integration into wider society	1.48 (1.02, 2.13)	1.50 (1.03, 2.18)

[a]Participants who reported that their health was fair, bad or very bad were defined as having poor self-rated health.
[b]Participants who scored 4 or more on the 12-item General Health Questionnaire were defined as having poor mental health.

Whatever hazards the physical environment may pose to human health should not deflect attention from the social and economic environment. These, too, pose risks for health.

REFERENCES

Brockes, E. (2001) "The phoney graffiti war and the killing that shocked an estate," *Guardian*, September 5, 2001.

Davey, S. G., Hart, C., Watt, G., Hole, D. and Hawthorne, V. (1998) "Individual social class, area-based deprivation, cardiovascular disease risk factors, and mortality: the Renfrew and Paisley study," *J. Epid. Comm. Health* **52**, 399–405.

Diez-Roux, A. V., Nieto, F. J., Muntaner, C., Tyroler, H. A., Comstock, G. W. and Shahar, E. (1997) "Neighborhood environments and coronary heart disease: a multilevel analysis," *Am. J. Epidemiol.* **146**, 48–63.

Diez-Roux, A. V., Merkin, S. S., Arnett, D., Chambless, L., Massing, M. and Nieto, F. J. (2001) "Neighborhood of residence and incidence of coronary heart disease," *N. Engl. J. Med.* **345**, 99–106.

Folsom, A., Stevens, J., Schveiner, P. J., McGovern, P. G. (1998) Body Mass index, waist/hip ratio, and coronary heart disease incidence in African Americans and whites. Atherosclerosis Risk in Communities Study Investigators. *Am. J. Epidemiol.* **148**, 1187–1194.

Fitzpatrich, J. and Jacobsen, B. (2001) Mapping health inequalities across London. London: The London Health Observatory.

Marmot, M. (2002) "The influence of income on health: views of an epidemiologist," *Health Affairs* **21**, 31–46.

Murray, C. J. L., Michaud, C. M., McKenna, M. T. and Marks, J. S. (1998) "US patterns of mortality by county and race: 1965–1994," pp. 1–97, Cambridge, MA, Harvard Center for Population and Development Studies.

Office for National Statistics (2001) *Geographic Variations in Health*, London: The Stationery Office.

Rowntree, B. S. (2001) "Poverty: a study of town life (1901)," In Davey Smith, G., Dorling, D., Shaw, M., (eds.). *Poverty, Inequality and Health in Britain, 1800–2000: A Reader*, pp. 97–106, Bristol: The Policy Press.

Stafford, M., Bartley, M., Mitchell, R. and Marmot, M. (2001) "Characteristics of individuals and characteristics of areas: investigating their influence on health in the Whitehall II study," *Health and Place* **7**, 117–129.

Stafford, M., Bartley, M., Wilkinson, R., Boreham, R., Thomas, R., Sacker, A. and Marmot, M. (2003) "Measuring the social environment: social cohesion and material deprivation in English and Scottish neighbourhoods," *Environ. Planning* **A35**, 1459–1475.

Townsend, P., Phillimore, P. and Beattie, A. (1988) *Health and Deprivation: Inequality in the North*, London, Crom Helm.

Chapter 7

The Air over London

Helen ApSimon

INTRODUCTION

It is almost 50 years since the Great Smog of 1952, when between December 5th and 9th the concentration of sulphur dioxide and total particulate matter rose to deadly levels between 1000 and 2000 micrograms per cubic metre or $\mu g.m^{-3}$. This was the most extreme example in a sequence over many years since the 19th century of 'pea soupers'; so called because of the characteristic grey green colour of the dense fog that brought visibility down to little more than a metre. In that one event it was estimated that between three and four thousand deaths occurred prematurely, particularly attacking the weaker and more vulnerable. This led to the Clean Air Act in 1956, with its extensive legal powers given to national and local agencies of government to curb the emissions from domestic and industrial coal burning, and to take measures to minimise the impact of energy generation and power stations. The result was a 50% decrease in sulphur dioxide (SO_2) concentrations and 70% in air borne particles emitted from fossil fuel combustion in the UK. But policies to reduce SO_2 concentrations in the UK such as the building of taller chimneys to disperse and dilute pollutants did not help solve the trans-boundary air pollution problems of acid rain.

London now produces different types of atmospheric pollutants that affect human health and plant life. These are largely induced by the emissions of road traffic. Episodes of high air pollution still occur, though they are different in nature from the former smogs. London with a prosperous, energy consuming population of over 7 million still suffers from poor air quality along with several other major European cities. Milan for example, although much smaller with only 1.4 million people, is sited in the populated Lombardy region, enclosed by mountains where stagnant air masses produce photochemical smog in summer similar to that in Los Angeles. Strategies there are currently directed to controlling emissions during episodes, for example by banning some of the traffic when air quality deteriorates. Fortunately the London region is less conducive to these extreme situations, but its sheer size and location between low hills and near an estuary cause its own unique problems.

Regular monitoring shows that in both central and suburban areas of London the concentration of pollutants regularly exceed the levels specified by the EC and the UK government for protection of human health. This chapter illustrates the research undertaken to observe and analyse the particular situation of London, and the modelling studies used to assess future situations. We also ask to what extent new technologies might enable air quality to improve over the next 10 years.

The chapter draws on work by members of the APRIL (Air Pollution Research in London) research network, originally founded by the Engineering and Physical Sciences Research Council, EPSRC and with additional support from the Environment Agency and DEFRA. The purpose of establishing APRIL has been to bring together the research community with those responsible for air quality in London, and to develop appropriate research to answer critical questions.

AIR QUALITY OBJECTIVES

The setting of air quality objectives is intended to improve the protection of human health, drawing on clinical and epidemiological evidence and the expertise of both international bodies such as the World Health Organisation (WHO), and, in the UK, the Expert Panel on Air Quality Standards. Currently 7 pollutants are targeted in European Commission (EC) Directives and in the UK Air Quality Strategy for the purposes of local air quality management (see Table 1). Of these 7 pollutants the two that widely exceed these objectives are nitrogen dioxide (NO_2) and PM_{10} (fine particles with aerodynamic diameter less than 10 microns, and hence small enough to penetrate to the lungs). Both of these are largely attributable to

Table 1 Air Quality Objectives for the protection of human health.

Pollutant	Objective	Target date
Benzene	16.25 μg.m^{-3} (5 ppb) annual mean	end 2003
1,3 butadiene	2.25 μg.m^{-3} (1 ppb) annual mean	end 2003
Carbon monoxide	11.6 mg.m^{-3} (10 ppm) 8 hour mean	end 2003
Lead	0.5 μg.m^{-3} annual mean	end of 2005 and half by 2008
Nitrogen dioxide	200 μg.m^{-3} (105 ppb) 1 hour mean (exceed less than 18 times/yr)	end 2005
	40 μg.m^{-3} (21 ppb) annual mean	end 2005
Particles (PM$_{10}$)	50 μg.m^{-3} daily mean (exceed less than 35 times/yr)	end 2004
	40 μg.m^{-3} annual mean	end 2004
Sulphur dioxide	350 μg.m^{-3} (132 ppb) 1 hour mean (exceed less than 24 times/yr)	end 2004
	125 μg.m^{-3} (47 ppb) daily mean (exceed less than 3 times/yr)	end 2004
	266 μg.m^{-3} (100 ppb) 15 minute (exceed less than 35 times/yr)	end 2005

ppb/ppm: parts per billion/million

traffic, and are a major focus in this chapter. Other pollutants such as SO_2 and lead have been greatly reduced, the latter with the introduction of lead-free petrol. In some cases the standards are based on long-term exposure over many years, for example in the case of 1,3 butadiene where the concern is cumulative risk of cancer; other standards are based on episodic peak concentrations, for example carbon monoxide (CO) which affects haemoglobin and the oxygen carrying capacity of the blood. However, the two pollutants that are the most difficult to control are NO_2 and PM$_{10}$. Limits are set for both short term (e.g. daily) and annual average concentrations. For PM$_{10}$ the large day to day variability means that if there are less than 35 days per year when concentrations of PM$_{10}$ are above 50 μg.m^{-3}, the limit of 40 μg.m^{-3} for the annual average will be achieved automatically: hence it is the first that needs to be addressed. However for NO_2 the target for annual average is more difficult to achieve than for the short term average. (This could change as the standards are tightened in future.)

The task of assessing air quality management areas, AQMAs, and the preparation of action plans for improvement of air quality, has been devolved to local authorities. In London this requires a close collaboration between London Boroughs, and the Greater London Authority (GLA), guided by the Mayor's transport and air quality strategies (see Chapter 5). Because of its particular problems some relaxation of the future objectives and target dates has been agreed for London.

OBSERVATION AND MEASUREMENT OF AIR POLLUTION OVER LONDON

Essential to the assessment of air quality is the assembly of reliable measurements collected from a network of monitoring stations. Many of the measurements of key pollutants come from the extensive London Air Quality Network, LAQN. These data are coordinated by the Environmental Research Group (ERG) at King's College London, who also under take quality assurance to ensure consistency and accuracy. Together with other networks managed and operated by ERG covering neighbouring areas of Surrey, Essex, Hertfordshire, Bedfordshire and Kent, these networks comprise over 110 monitoring sites. Further information on the network and measurements made can be found on the website (www.erg.kcl.ac.uk/home.asp).

An example of the way the ERG have carefully analysed the data to reveal trends in air pollution is illustrated in Figs. 1(a) and 1(b). The first figure, 1(a), shows a clear downward trend in the average concentrations of SO_2, CO and PM_{10} since 1997, reflecting cleaner fuels, and improvements in vehicle technology and exhaust emissions. This is a significant indicator of improvement. Figure 1(b) shows that despite a decrease in NO_x, there is less of a trend in the pollutant NO_2. NO_x is a more complex pollutant in that the oxidised nitrogen is emitted mainly as nitric oxide NO with only a small fraction in the critical form of nitrogen dioxide, NO_2. However further NO_2 is formed in the atmosphere by a chemical reaction between NO and atmospheric ozone, O_3. This dual origin of NO_2 leads to a complex and variable relationship across the area of London between the concentrations of NO_2 and NO_x, and complicates the assessment of air quality management areas, as discussed further below. The amount of secondary NO_2 formed is limited by the availability of the ozone, which is depleted in the process (thus ozone tends to be lower in polluted urban areas than in surrounding rural areas, contrary to many peoples'

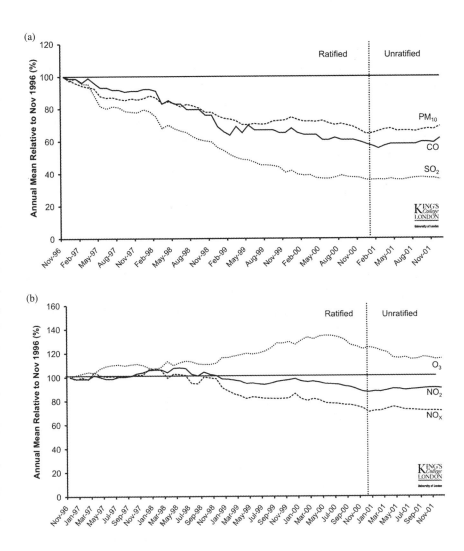

Fig. 1 (a) Trends in CO, PM_{10} and SO_2 over London (King's College London based on LAQN data). (b) Corresponding trends in NO_x, NO_2 and ozone, O_3.

expectations). The supply of ozone in the background air approaching London is the result of photochemical processes over large areas of Europe and its nearby ocean, mainly involving hydrocarbons and NO_x. Because background ozone concentrations appear to be increasing at up to 1 ppb per year, it is more difficult to reduce NO_2 concentrations, for example by reducing traffic or introducing new technologies. A substantial reduction in emission of NO_x is required to attain the air quality objective for NO_x in London by 2010, and some localised exceedance is likely (DEFRA 2004).

In addition to routine monitoring of targeted pollutants, a wider range of measurements is needed to cover other atmospheric contaminants and to explore, for example, the size and composition of fine particles. Many such measurements are undertaken in parallel with routine monitoring at the DEFRA "supersite" in Marylebone Road (visible as a green cabin on the pavement on the opposite side of the road from Madame Tussaud), which is also managed by King's College. Detailed inter-comparison of different techniques is also important. Thus it has been recognised that the routine equipment used to monitor fine particles loses the more volatile components, and hence does not register the whole mass. This loss varies with the chemical composition of the particles, but is on average about one quarter of the overall mass, resulting in a compensation factor applied to the raw monitoring data before making comparison with the air quality objectives.

All the measurements above are made at street level. However it is already clear that the atmosphere above London influences air quality at the ground. Higher altitude measurements above the urban canopy of buildings could help the analysis of London's air pollution problems, for example deploying powerful remote sensing techniques such as lidar to observe vertical profiles of ozone and fine particles, or even on tall buildings. For example measurements of particulates are taken on the top of the BT tower (200 m) by Dr. Steve Smith and colleagues at King's College, London.

Satellite data also provides useful spatial and temporal records of air pollution. Work by Doyle and Dorling (2001), for example, has illustrated how such data can reveal broad bands of pollution stretching across the UK from northern France and continental Europe, giving a large scale picture which is difficult to assemble from ground based monitoring networks. Modern satellite technology provides vertical profiling of the lower atmosphere down to the ground, giving a three dimensional distribution of aerosols and trace gases throughout the troposphere. The technology can also be used to measure aerosol amounts over both ocean and land surfaces, including large conurbations. Such capabilities could become very useful in assessing both particles and ozone advected to London from the Continent in anticyclonic conditions, contribute appreciably which to London's air quality problems.

WHAT CAUSES HIGH LEVELS OF POLLUTION OVER LONDON?

The high pollutant concentrations in London are largely the result of a high density of emissions generated over this large area. But, as indicated above, meteorology and the air imported into London are also important factors. Polluted air from continental Europe tends to have high burdens of 'secondary' particulate material, containing sulphate and nitrate, resulting from oxidation in the atmosphere of sulphur dioxide and nitrogen oxides. In spring and summer, the incoming air also has high ozone levels. This enhances the capacity for oxidation of fresh emissions over London of nitric oxide, NO, to form nitrogen dioxide NO_2, and also nitrate aerosol. The production of NO_2 in the atmosphere means that reducing emissions of nitrogen oxides by vehicles can only lead to a smaller proportional decrease in NO_2. This is why emissions have to be reduced by an even greater factor to improve air quality significantly.

Meteorological conditions determine the rate of dilution of pollutants as well as affecting the chemical process. Thus with stagnant air masses, there is very little dilution, so that pollutants can accumulate. Severe pollution episodes arise in these situations when there are low level inversions which restrict mixing of emissions to a shallow layer of air only one or two hundred metres deep. Such episodes are more frequent in winter, as occurred during the severe episode in 1991, a modern version of the former smogs. The levels of SO_2 were low, but extreme concentrations of NO_x of around 1000 ppb were a significant health hazard. To reduce the possibility of future episodes, there have been suggestions for new regulations in London comparable to those in other European cities, where traffic and industrial activity are reduced until the pollution concentrations subside.

THE METEOROLOGY OF LONDON

Because of its size London has a major influence on its own meteorology. In urban areas there are many more obstacles to the airflow and they are also higher than trees and hedges in the surrounding rural areas. The surfaces of roads and buildings also greatly affect the transfer of heat and water between the ground and the atmosphere. These factors affect temperatures and turbulence in the 'mixed layer' of air within one or two hundred metres of the ground on a cold still night or up to one or two kilometres on a sunny day. The orography of the Thames valley also has the effect of channelling east-west winds (which is why London used to have many windmills (Aykroyd 2001)). The proximity of the coast and estuary can cause complex shallow sea-breeze wind patterns differing from the larger scale air flow over southeast England (Chandler 1965).

Recent research in atmospheric science and numerical simulation techniques has led to great advances in the accuracy of prediction of the meteorological situation over London and the computation of the details of local atmospheric (or 'mesoscale') flows builds on the same methods as for weather prediction, but with finer detail. The smallest scale is now down to 1 km (for research purposes). Such models could be used to predict transport of pollutants as well as the meteorological flow, and in due course air pollution chemistry.

As an example, the model of the Joint Centre for Mesoscale Meteorology, involving the Met Office and the University of Reading has been run for a 'typical' summertime urban heat island case study (10th to 11th May 2001) to examine the impact of urbanisation. Land use data from the Centre of Ecology and Hydrology (CEH), has been used to specify the proportion of different surface types, including 'urban' in each grid square. In this case, a substantial daytime and night time urban heat island develops. The impact of the urban areas on the meteorology is demonstrated in a comparison between the calculation of atmospheric temperature when the urban surfaces are replaced by vegetation. Figure 2(a) shows the resulting

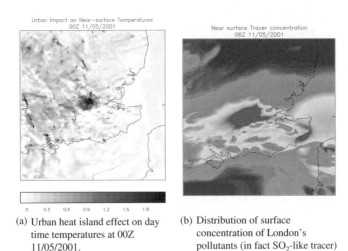

(a) Urban heat island effect on day time temperatures at 00Z 11/05/2001.

(b) Distribution of surface concentration of London's pollutants (in fact SO_2-like tracer) at dawn (06Z 11/05/2001).

Fig. 2 Special features for the environment of London illustrated by very small scale computations (with resolution of 1 km) using the Met Office 'mesoscale' numerical weather prediction model.

increase in temperature over London and downwind areas. This warming also increases the vertical dispersion of pollution within the mixed layer.

Figure 2(b) shows how the complex air flow over the London basin affects the surface concentration of a SO_2-like tracer at dawn. Most of the flow is along the Thames Valley, with little transport over the North Downs, except where material 'leaks' through valleys.

MODELLING AND ASSESSMENT OF POLLUTANT CONCENTRATIONS AGAINST OBJECTIVES

To make detailed calculations of acute episodes of high air pollution over London, it is necessary to combine advanced models of mesoscale meteorology and atmospheric chemistry, an approach which is now common practice in the USA. Such detailed calculations take many hours with the fastest computers and therefore are not practical for daily forecasts of pollution.

The main practical reason for calculating pollutant concentrations is to assess how seriously they exceed air quality objectives, and for making future projections based on different 'scenarios' for emissions and for the development of the urban area. Since many calculations are needed for all the meteorological situations and scenarios involved, simple and fast modelling methods are needed. A basic requirement for all such models, simple or complex, is a good emissions inventory, describing the magnitude, spatial distribution and temporal variation of emissions of relevant pollutants.

A substantial inventory of emissions for London now exists. This has been coordinated by the GLA, and includes very detailed data on traffic, with vehicle flows and speeds on individual major roads derived from census data and traffic modelling, combined with background contributions from minor roads and other sources ascribed to a grid of 1 times 1 km grid squares. Different emission factors are defined for different classes of vehicle according to their speed and age, based on measurements of existing vehicles exhaust emissions and projections of improvements in new vehicles in accordance with agreed emission standards. Additional contributions from cold start conditions, and from brakes and tyres are also added (GLA, http://www.london.gov.uk/mayor/air_quality/model. jsp#atmosphere).

Modelling studies have used the emission data to map concentrations across London both for the present situation, and to predict areas where air quality objectives are still likely to be exceeded in 2004 if no further measures are taken. Such calculations provide a basis for declarations of Air Quality Management Areas by local authorities in accordance with legislation. Two types of modelling approach have been widely used in London. The first, by David Carslaw and colleagues at King's College (Carslaw et al. 2001), is an empirical statistical approach whereby background concentrations of NO_x and PM_{10} are correlated with emissions from local grid squares. Then the contribution of individual major roads is superimposed link by link. The mapping of NO_2 is derived from the mapping of NO_x, assuming a defined relationship between NO_x and NO_2 which reflects the dependence of oxidation by ozone of NO to NO_2.

The second approach makes an explicit, but approximate, calculation of the dispersal processes and atmospheric transport of the pollutants directly for each meteorological and emissions scenario, taking into account all sources within the emission inventory area (in the case of London the area covered by the London Atmospheric Emission Inventory). One such model that has been widely tested and applied in London is that of Carruthers et al. (2000) at CERC Ltd, ADMS 3*. This model allows for various types of emission and the complex dispersion processes in the urban environment down to the scale of the street (i.e. much smaller than the meteorological 'mesoscale' approach). Thus a "street canyon" sub-model is included to estimate concentrations at street level where dispersion is affected by buildings and vehicle induced turbulence. The ADMS-Urban model is run using successive hours of meteorological data, background pollution data and emissions data as input. Both long term averages, shorter averaging times (as little as one hour) and percentiles can be calculated. Air quality from local sources is affected by the most recent meteorology while other sources are affected by current meteorology and that prevailing over previous hours, which determines how they have been transformed by chemical reactions.

The map in Fig. 3(a) shows annual average concentrations of NO_2 in $\mu g.m^{-3}$ for 1999. This indicates clearly that there are areas in central and west London, and along major roads which are currently in excess of the standard of $40 \mu g.m^{-3}$ (21 ppb).

The fluxes of pollutants entering the London area have to be estimated from rural measurements, or from larger scale models. Calculations of the

---

*ADMS 3 Technical Specification (2000) Cambridge Environmental Research Consultants, http://www.cerc.co.uk/software/publications.htm.

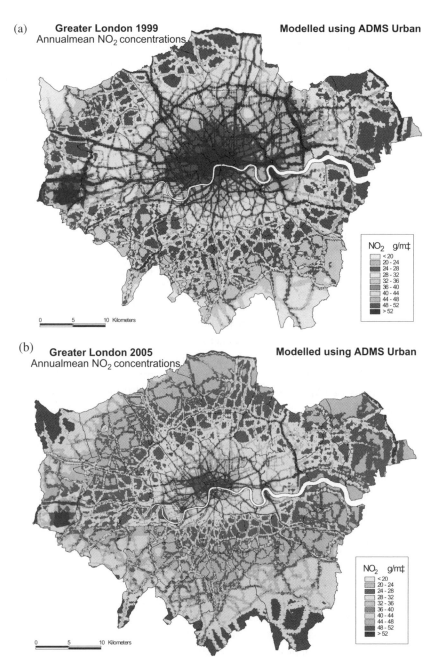

Fig. 3 Map of annual mean NO_2 concentrations at ground level over London modelled using ADMS-Urban for (a) 1995 and predicted for (b) 2005.

concentration of smallest particulates (PM_{10}), show that they have a similar distribution across London as that of pollutant gases such as NO_2. The highest levels are shown to be close to roads with maximum traffic densities. New models for some of the more complex features of urban air flow and dispersion, such as the occurrence of high concentrations at congested road junctions, are being developed in the inter university research programme DAPPLE (www.dapple.org.uk) supported by EPSRC.

Around Heathrow Airport to the west there is the additional contribution to the pollution levels from aircraft emissions. Although the emission inventory for Heathrow does indeed indicate that aircraft produce the largest proportion of emissions, in the airport area, they mainly occur during take off and landing of aircraft on sloping flight paths. Since a large proportion of these emissions are therefore generated above 100 metres, they are diluted before reaching ground level; as are industrial emissions from tall chimneys. Studies of concentrations of NO_x round Heathrow airport at Imperial College, show that emissions from vehicular traffic in the airport area and its surroundings contribute much more than the aircraft to ground level concentrations of air pollution.

A similar picture to that for NO_2 across London emerges for PM_{10}, showing the highest levels close to roads where there are the greatest traffic densities. Figure 4 illustrates how different sources contribute to annual average concentrations of PM_{10}, taken from work at Imperial College on abatement strategies. Here various combinations of technological measures such as particle traps, and traffic control measures are assessed using atmospheric modelling (Mediavilla-Sahagun and ApSimon 2003). Each column gives a break down of concentrations of PM_{10} at a busy road side site in one of a row of 1 km grid squares from west to east across London. The top section of each column indicates the contribution from traffic in the local road, superimposed on the background contribution from other primary sources of PM_{10} in the London inventory, subdivided into contributions from central, inner and outer London, and the M 25. However below this there are two further important contributions across the whole of London, not associated with sources in the London inventory. The broader upper band corresponds to the secondary particulate contribution formed by chemical oxidation of pollutants to form sulphate and nitrate, much of it imported from Europe. This contribution to particulate exposure should reduce in future in accordance with emission reductions agreed under the Gothenburg protocol (www.unece.org/env/lrtap). But the reduction will be highly variable — emphasising the need to have a good understanding of episodes and the appropriate research tools as discussed above.

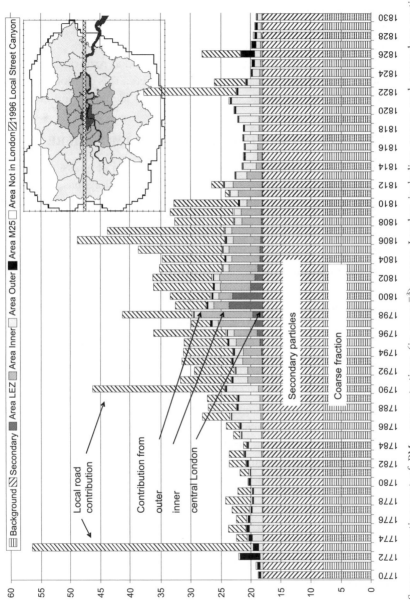

Fig. 4 Source apportionment of PM_{10} concentrations (in $\mu g.m^{-3}$) across London including coarse and secondary particulate contributions (1996). (From Urban Scale Integrated Assessment Model, USIAM (Mediavilla-Sahagun and ApSimon 2003).)

The final contribution represents the contribution of coarser particles in the range between 2.5 and 10 microns. Most emissions from vehicles are finer than this (indeed exhaust emissions are often as ultra-fine particles which some health experts believe are particularly to blame for observed health effects). But the sources of the coarser fraction are still not well understood, with various suggestions of re-suspended road dust, erosion of road surfaces, building and construction sites, etc. This portion cannot be represented properly until the sources can be fully identified and quantified.

FUTURE IMPROVEMENTS IN LONDON'S ATMOSPHERE

Since local traffic is responsible for a large proportion of emissions and this can be controlled by local action, this source of emissions has been particularly targeted for improving London's air quality (see Chapter 5). Control of domestic and industrial emissions also play a part, as well as reductions in trans-boundary air pollution and imported particulates. But as we have explained, the rate at which the locally produced nitrogen oxide NO is converted into the more noxious nitrogen dioxide NO_2 depends on ozone concentrations in air entering London, and this depends on total hydrocarbon and oxides of nitrogen emitted on a global scale. This is why control of air pollution levels over London depends on a combination of local, European and international actions to reduce emissions.

We should note what progress is underway. Previous action on tight standards is already leading to vehicles being better maintained, and the introduction of new types of vehicles with significantly reduced emissions. The Mayor's Transport strategy (see Chapter 5) also targets vehicle emissions, with the aims of reducing-traffic, encouraging low emission vehicles, especially in central London, and a greater use of public transport. Figure 3(b) shows the projected map of NO_2 concentrations in 2005, indicating the improvement anticipated by direct comparison with the map for 1999. This still indicates areas of exceedance, requiring further emission reductions before the air quality objectives can be met.

In considering future possible scenarios up to 2020, the introduction of new technology can help to improve London's air quality further, such as the use of alternative fuels including bio-fuels, electric or hybrid electric-combustion vehicles, and fuel cells using hydrogen. The latter generates only water and a very small amount of NO_x. Emissions of CO_2 by vehicles which contribute about 25% to greenhouse gas emissions also have to be reduced. This requires an integrated approach to improving local environment and mitigating local climate change. Scenarios for emissions from traffic explored by Marta Iglesias (2004) imply a further reduction in NO_x emissions for a 'business as usual scenario' of about 65%. The use of new technologies mentioned above provide scope for further reduction in both NO_x and PM_{10} emissions as well as reducing CO_2 and other pollutants. But it is important especially in the case of fuel cells to look at the overall picture including the emissions produced in the generation of the hydrogen to fuel the vehicles. Of course the overall improvement in London also depends on how much the industrial and building related emissions can be reduced.

CONCLUSION

As the understanding of the health effects of different pollutants develops, particularly fine particles, the air quality objectives are likely to change as time passes. At the same time research on the atmospheric processes, new developments in automobiles, energy and building, and policy need to be combined to develop strategies to meet those objectives. New prediction and control approaches maybe necessary to understand and then to prevent episodes of poor air quality over London, and to relate emission reduction strategies in London with continental and global initiatives.

ACKNOWLEDGEMENT

I am very grateful as Chairman of the APRIL (Air Pollution Research in London) network for the contributions from King's College London (Gary Fuller, David Carslaw and Steve Smith): David Carruthers and Jo Blair at CERC; the Met Office (Peter Clark, Doug Middleton and Dick Derwent), University College London (Peter Mueller), University of Hertfordshire (Ranjeet Sokhi), University of East Anglia (Martin Doyle and Steve Dorling), University of Surrey (Alan Robins) and staff (Roy Colvile, Kevin Clemitshaw, Mike Jenkin) and students at Imperial College (Antonio Mediavilla, Fernando Farias, Kiki Asimakopoulos, and Marta Iglesias) and last but not least the APRIL coordinator Linda Davies.

REFERENCES

Aykroyd, P. (2001) *London — A Biography*, Viking, London.
Carruthers, D. J., Edmunds, H. A., Lester, A. E., McHugh, C. A. and Single, R. S. (2000) "Use and validation of ADMS-Urban in contrasting urban and industrial locations," *Int. J. Environment and Pollution* **14**, pp. 1–6.

Carslaw, D., Beevers, S. and Fuller, G. (2001) "An empirical approach for the prediction of annual mean nitrogen dioxide concentrations in London," *Atmospheric Environment* **35**, pp. 1505–1515.

Chandler (1965) *The Climate of London*, Hutchinson, London.

DEFRA (2004) "Nitrogen dioxide in the United Kingdom," *First Report of the Air Quality Expert Group*, AQEG.

Doyle, M. and Dorling, S. R. (2001) "Satellite based monitoring of aerosol plumes," *Proceedings of the 3rd International Conference on Urban Air Quality*, Loutraki, Greece.

Iglesias, M. and ApSimon, H. M. (2004) "Alternative vehicle technologies and fuels in scenarios for atmospheric emissions in London," *J. Environmental Assessment Policy and Management*.

Mediavilla-Sahagun, A. and ApSimon, H. M. (2003) "Urban scale integrated assessment of options to reduce PM_{10} in London towards attainment of air quality objectives," *Atmospheric Environment* **37**, pp. 4651–4665.

Chapter 8

The Art and Science of London's Atmosphere around 1900

Dr John E Thornes and Gemma Metherell

A great chocolate-coloured pall lowered over heaven, but the wind was continually charging and routing these embattled vapours, so that as the cab crawled from street to street, Mr Utterson beheld a marvellous number of degrees and hues of twilight; for here it would be a glow of a rich, lurid brown, like the light of some strange conflagration, and here, for a moment, the fog would be quite broken up and a haggard shaft of daylight would glance in between the swirling wreaths.

(Robert Louis Stevenson: Doctor Jekyll and Mr Hyde 1886).

The history of the scientific and artistic studies of air pollution and urban atmosphere sheds light on current concerns about how climate change and the development of cities might affect London's air quality over the coming decades. Smoke pollution and the permanent smoke haze that enveloped the Victorian cities of Britain not only ruined people's health through rickets, bronchitis, pneumonia and asthma but it also altered the climate, reducing sunshine and increasing the number and severity of fog episodes. The climate in these cities in the 19th century was far more dramatic and visual than it is today as smoke blotted out the sky and also gave the air a filthy smell and taste. The horrors of smoke pollution and the growth of legislation to try and deal with it have been well documented, not just in Britain (Brimblecombe 1988; Mosley 2001; Luckin 2002) but also in Germany (Brüggemeier 1994) and the United States (Stradling 1999).

These historical studies of urban atmospheres are part of a wider debate about air pollution and society and even more broadly about weather, climate and society (Janković 2000; Hamblyn 2001). A new subject of 'cultural climatology' is emerging (Thornes and McGregor 2003)[a] built in part on earlier 'deconstructing' studies of weather, climate and air pollution in landscape painting (Thornes 1979, 1999; Brimblecombe 2000; Brimblecombe and Ogden 1977; Gedzelman 1991) which showed the symbolic significance of atmospheric effects. Bonacina (1939:485) defined 'landscape meteorology' to be 'those scenic influences of sky, atmosphere, weather and climate which form part of our natural human environment ... whether the natural scenery is merely received and carefully stored in the memory, or is photographed, painted or described.'

This chapter is concerned with the relationship between the climate of London and the remarkable impressionist paintings by the French artist Monet at the turn of the 20th century. Monet's 'London Series' will be deconstructed in the light of the uncertain knowledge of weather and climate at that time and also the prevailing artistic and literary culture. A fundamental question is whether Monet's images of London at this time represent reality or are they figments of the artist's imagination or a combination of the two? The distinction between art and nature is never clear. The term 'Realism' was coined in France with respect to the French artist Gustave Courbet (1819–1877) who produced 'The Realist Manifesto' in 1855 (Nochlin 1966). Baudelaire stated that the Realists "want to represent things as they are, or as they would be, supposing that I (the perceiving subject) did not exist" (Quoted in Rubin 1996:53). However Realism in art developed beyond just a representation of real and existing things, it became a movement to overturn the established view of art and paved the

---

[a]An earlier version of this chapter was published in *Weather, Climate, Culture* edited by Sarah Strauss and Ben Orlove, Chapter 8 Monet's 'London Series' and the Cultural Climate of London at the Turn of the 20th Century, pp. 141–160, Berg Press, Oxford 2003. Thornes and McGregor (2003:178) discuss the concept of Cultural Climatology as follows: we advocate that climatology should not only be concerned with the study of physical processes at various space and time scales but with evaluating and understanding climate society interactions and feedbacks as manifest by societal response and how society may interpret climate information. Therefore, we view climate as an integral part of culture and as such we contend that there is a need to develop a sub-discipline within climatology that we will refer to as cultural climatology.

way for the Impressionist movement by hastening the departure of the hitherto dominant 'classical' school. Realism attempted 'to create objective representations of the external world based on the impartial observation of contemporary life' (Rubin 1996:53). Realists could therefore paint their own vision of nature although their observations were rarely impartial and the movement became closely associated with wider socio-political views. As a result the term 'Naturalism' arose to describe art without a particular socio-political significance. In its broadest sense Naturalism refers to any work of art that depicts actual rather than imaginary or exaggerated subject matter. The subject that is represented by the artist is done so as naturalistically as possible, without deliberate idealisation or stylisation. In England we would describe Turner as a realist and Constable as a naturalist, despite the fact that both artists had died well before the terms were defined. Turner strove to incorporate his own emotions and ideas into his art whereas Constable found his art under every hedge and within every cloud. Can Monet be best described as an exponent of realism or of naturalism or both?

English weather has always attracted the attention of artists due to its transitory nature that constantly challenges their ability to catch the 'atmosphere' on canvas.[b] As the weather and climate of England has changed over the centuries, so this is reflected in the art of the times, which offer unique visual 'weather diaries' of transition. Lamb (1967) and Neuberger (1970)[c] have shown, by examining landscape paintings from around the

---

[b]The landscape artist Turner (1775–1851) spoke of the advantages of the British climate for landscape artists:

> *In our variable climate where [all] the seasons are recognizable in one day, where all the vapoury turbulence involves the face of things, where nature seems to sport in all her dignity and dispensing incidents for the artist's study... how happily is the landscape painter situated, how roused by every change in nature in every moment, that allows no languor even in her effects which she places before him, and demands most peremptorily every moment his admiration and investigation, to store his mind with every change of time and place, (Wilton 1979:107).*

[c]Lamb examined 200 Dutch and British paintings from 1550 to 1939. Neuberger examined 12 284 paintings in 41 art museums in 17 cities of 9 countries. He examined directly the weather in the paintings (clouds, visibility etc.) and indirectly the clothing of the people depicted to give an indication of the season etc.

world, that artists have faithfully reproduced the changes of climate in their works. For example the works of Breugel depict the 'Little Ice Age' which led to the regular freezing of lakes and canals and rivers such as the Thames in winter. Obviously one has to be aware that artists may freely use their 'artistic license' to exaggerate and invent. This makes their study even more fascinating as we attempt to unravel the embedded culture, as well as the realities of the weather and climate in their paintings. Pevsner (1955:20)[d] in his classic book *The Englishness of English Art* suggests that although the English loved to complain about the weather at that time, they did nothing about it: 'Perhaps this staunch conservatism in the teeth of the greatest discomforts is English?' The 'stiff upper lip' and indifference to the domestic and industrial pollution in England suggested that the London Fog was not bad enough to urge much social or political action even though 200 years earlier regulations had been introduced regarding the burning of coal (Chapter 13). The London Particular was ignored by all except the patrons of the arts:

> *At present, people see fogs, not because there are fogs, but because poets and painters have taught them the mysterious loveliness of such effects. There may have been fogs for centuries in London. But ... They did not exist till Art had invented them. (Wilde 1889:925)[e]*

It is to foreign visitors that we owe the description of the worsening combination of smoke and fog and the changing English urban climate as the 19th century progressed.

The most distinguished of these was Monet whose fascination with the London's foggy atmosphere resulted in his London Series of 95 paintings. These can best be understood by closer study of the pictures themselves and by examining the actual sites of the images and where they were

---

[d]The first chapter of Pevsner's book is called 'The Geography of Art' in which he gives a 'whole string of facts from art and literature tentatively derived from climate.' (p. 19)

[e]Oscar Wilde (1889:925) suggested that nature imitates art:

> *Where, if not from the impressionists, do we get those wonderful brown fogs that come creeping down our streets, blurring the gas lamps and changing the houses into monstrous shadows? ... The extraordinary change that has taken place in the climate of London during the last ten years is entirely due to a particular school of art.*

painted (Rose 2000)[f]. It is ironic that we must turn to a Frenchman for the visual representation of the London Particular and that none of Monet's London Series are on permanent display in London galleries today. Perhaps the English do not want to be reminded that 'Hell is a city much like London — a populous and a smoky place'.[g]

The climate and culture of London at the end of the 19th century represent a fascinating enigma. London was the imperial metropolis of the world and the capital of a British Empire with a population of four hundred million people, the biggest Empire the planet had yet seen (Schneer 1999). London's dominant culture in the Empire was exported to the rest of the world. However there was a hefty price to pay for all this power, industrial production and trade. Although the fact that the buildings and gardens of cities causes their climate to differ from that of the surrounding countryside was well recognised even in the Ancient World, London was the first city in which the 'urban climate' was measured (Howard 1818). As the most polluted city in the world, London had become affectionately known around the globe as 'The Big Smoke' and the word smog (a combination of smoke and fog) was first coined in 1905 with reference to the London fog. This was one product that could not be exported to some distant colony; London and its many visitors had to live with it. Ruskin (1884) in 'Storm Cloud of the Nineteenth Century' notes his diary entry for Tuesday 20th February 1872:

> *There has been so much black east wind lately, and so much fog and artificial gloom, besides, that I find it is actually some two years since I last saw a noble cumulus cloud under full light. I chanced to be standing under the Victoria Tower at Westminster, when the largest mass of them floated past... and I was more impressed than ever yet by the awfulness of the cloud-form, and its unaccountableness, in the present state of our knowledge. The*

---

[f]Rose's book is an excellent introduction to the critical interpretation of visual images. She suggests a methodological framework that involves the examination of the site of production, the site of the image itself and the site of audiencing. In order to try and understand the different, sometimes controversial, approaches to these aspects, she defines three modalities: (1) Technological: 'any form of apparatus designed either to be looked at or to enhance natural vision, from oil painting to television and the internet.' (2) Compositional: 'when an image is made, it draws on a number of formal strategies: content, colour and spatial organisation...' (3) Social: 'the range of economic, social and political relations, institutions and practices that surround an image and through which it is seen and used.'

[g]Quoted from Shelley (1819).

Victoria Tower, seen against it, had no magnitude: it was like looking at
Mount Blanc over a lamp-post. (40)

Ruskin was convinced that he had discovered a new meteorological phenomenon to go with the fog:

This wind is the plague-wind of the eighth decade of years in the 19th cen-
tury; a period that will assuredly be recognised in future meteorological
history as one of phenomena hitherto unrecorded in the courses of
nature.... (43)

This plague-wind:

It looks partly as if it were made of poisonous smoke; very possibly it may be:
there are at least two hundred furnace chimneys in a square of two miles on
every side of me. But mere smoke would not blow to and fro in that wild way.
It looks more to me as if it were made of dead men's souls... (47)

Ruskin's claim that he had discovered a new type of 'plague' wind and
'storm' cloud was of course a myth — Athena raped by political economy
(Cosgrove and Thornes 1981). Ruskin was harking back to the pre-
modernist days when meteorology was concerned with classical 'meteoric
reportage' — when the atmosphere was totally 'unaccountable' and
unique. Also the skies all over the world had been richly coloured in 1883
as a result of particles spewed up into the atmosphere in the East Indies by
the eruption of the volcano Krakatoa.

In order to appreciate what was meant by meteorology in Ruskin's
time, we have to return to Aristotle and record that his ideas expressed in
'Meteorologica' had only been overturned in the last 100 years. Though
doubtless it was still influential among the majority of the British educated
classes brought up on classical literature. The important point (Janković
2000) was that the weather was not determined by 'meteorites' or 'meteors'
whose extra terrestrial origins were clarified around 1800. Thereafter the
study of meteors could be left to astronomers. In fact the study of the
atmosphere was passed over to chemistry and the genius of John Dalton,
who discovered the essential physics of atmospheric water vapour. As the
first chemical meteorologist he moved meteorology under the heading of
'chemical philosophy' along with chemistry and geology.

The atmosphere was then considered to be a vast chemical laboratory,
which suited the growth of laboratory science at the beginning of the 19th
century. It was not long before the chemical physicist John Tyndall realised
that the climate might be changed by the carbon dioxide emitted by human
activities.

Ruskin was following others back to classical times in seeking an understanding of the world and poetic and artistic inspiration from the weather. Writers in the 17th and 18th century had commented that "English air was the cause of the mutability of English thought and thus the source of national characteristics such as newfangledness, rashness and love of rebellion" (Janković 2000:3).

Generally the weather was reported in terms of *unusual* events at specific places — meteoric reportage. 'Meteoric' in the classical sense could include any usual weather events such as storms, earthquakes, fireballs, waterspouts, flying dragons or northern lights. Meteoric captures the unusual discrete meteorological events that are separated by 'anonymous' interludes of atmospheric tranquility.

This pre-1800 meteorology was a far cry from the content of modern meteorology. Meteorological phenomena were believed to explain bodily pain, death, financial loss and dreams as a scientific alternative to religious explanation in terms of divine intervention, sin and the devil. Meteoric events were publicised in pamphlets, almanacs, broadsheets, ballads, poetry, newspapers, drama, oral culture and even in the scientific journals of the time. Meteorology was about the unique, the unexplainable, the astonishing and was completely unfathomable.

Could a science of the weather be constructed amongst such unique events along the lines of Francis Bacon's precept that 'nature spoke more clearly when it sported itself in the "out of the ordinary"'? Perhaps the inductive collection of unusual events was leading to a better understanding through encouraging networks for the exchange of information (Janković 2000).

The London antiquarian Roger Gale, having received some letters from provincial correspondents such as clergymen and members of the gentry wrote 'who could have expected such a learned correspondence and so many curious observations … made by a set of virtuosi almost out of the world'! (quoted in Janković 2000:6).

Ruskin could not accept that science would ever explain how the atmosphere worked, especially as the success of astronomers in predicting future events such as eclipses was not mirrored by the meteorologists. Certainly there was no evidence of success in forecasting the onset of London Particulars. The modernist view that the behaviour of the atmosphere was predictable did not make significant strides until the beginning of the 20th century when Jakob Bjerknes in Germany and Norway led the way for Lewis Fry Richardson (1922) in England to publish *Weather Prediction by Numerical Process.*

Victorian London fog (the London Particular or London Peasouper) was probably the most famous global meteorological phenomena of the 19th and 20th century. This cultural enigma — a symbol of power, mystery and prosperity that defied the truth of the smoke poisoned lungs and badly bowed limbs due to rickets, caused by the lack of light, that afflicted tens of thousands of Londoners at any one time.

In the Victorian winter a million coal fires mixed smoke and sulphur dioxide with the industrial outpourings of a myriad of chimneys, furnaces, processing plants, railway engines, steam driven barges and boats on the Thames, to produce a London Particular more than 200 feet (60 m) thick. In 1873 it was noted that over 3 days in December there were up to 700 extra deaths, 19 of them as a result of people walking into the Thames, docks or the canals and drowning (Brimblecombe 1988:123).

Coal was first brought to London in appreciable quantities from Newcastle in the early 13th century being used as ballast for boats returning from Tyneside to London. By 1620 it was estimated that 100 000 tonnes per year were being imported and as supplies of wood dwindled so the sales of 'sea-coal' continued to rise. The first serious indictment against the deleterious effects of smoke from the running of sea-coal came in 1661 from John Evelyn in *Fumifugium*:

> *That this Glorious and Ancient City... should wrap her stately head in clouds of smoke and sulphur, so full of stink and darkness, I deplore with just Indignation.*

> *... the City of London resembles the face rather of Mount Etna, the Court of Vulcan, Stromboli or the suburbs of Hell, than an Assembly of Rational Creatures.*

The adverse effect of smoke mixed with accompanying sulphur dioxide was also noted by Evelyn:

> *For is there under Heaven such coughing and sniffing to be heard, as in the London churches and Assemblies of the People, where the barking and the spitting is most importunate.*

> *... but the chance for life in infants, who are confined in the present Foul Air of London, is so small, that it is highly prudent and commendable to remove them from it as early as possible.*

John Evelyn was a cultural climatologist centuries ahead of his time. He suggested moving industries downwind of the city and planting gardens along the Thames. His ideas were ignored and coal fired the industrial revolution in London and became the domestic fuel for all classes of society. By the middle of the 18th century a perpetual mist or fog enveloped London in winter and those sepia faded looking photographs of London in the second half of the 19th century show the perpetual mist not a fading photograph. In the summer months the sun did break through and the mist was temporarily lifted. On 3rd September 1802 Wordsworth wrote his poem *Composed Upon Westminster Bridge* in which he notes:

Ships, towers, domes, theatres and temples lie
Open unto the fields, and to the sky;
All bright and glittering in the smokeless air.

London was not oblivious to the smoke and fog. In 1819 a committee was appointed by Parliament to enquire 'how far persons using steam engines and furnaces could erect them in a manner less prejudicial to public health.' In 1853–1856 the Smoke Abatement Acts came into force relating to Metropolitan areas, however the impact on smoke levels was probably more imagined than real. In 1858 and 1866 the Sanitation Acts authorised sanitary authorities to take action against smoke nuisance but legislation did not seem to work. Smoke levels and the frequency of fogs continued to rise.

The relationship between smoke and fog was still not completely understood. There had always been fogs along the Thames in London and as the number increased, so people thought that it was obviously caused by the increasing smoke. There were no observations of smoke levels and meteorological observations recorded fogs but not their cause. The first serious study of the increase in fogs, based on proper observations rather than on conjecture, was published by Brodie (1892). He showed that there had been a steady increase in the prevalence of fog in London between 1870 and 1890.

However in 1905 he published a further article (Brodie 1905) stating:

In the 13 years which have since elapsed the tendency has been so strongly
in the opposite direction that little apology is needed for bringing the subject
once more before the notice of the society.

The fog frequency was recorded at the Meteorological Office official London site in Brixton, and data was presented for the years between 1871 and 1903. The highest total of 86 was recorded in 1886 and the lowest — just 13 in 1900, with a mean of 55. Brodie suggests a number of reason as to why the fog levels had fallen including the success of the Coal Smoke Abatement Society; modern grates and stoves that were more efficient at burning the smoke; introduction of incandescent gas and electric lights and the increasing use of gas stoves for heating and cooking.

Brodie's paper caused quite a stir at the meeting of the Royal Meteorological Society at which it was presented. The discussion was held over to the next meeting allowing the Fellows to prepare some statistics of their own. Mr Marriot said that Mr Brodie needed to define what he meant by a fog. He assumed that Mr Brodie's data related to what was popularly called 'London Fog' and not to ordinary meteorological fogs. Mr Marriot presented his own findings in comparison for days of fog at West Norwood some 3 miles from Brixton. The fog frequency for the 26 years of comparison averages 57 days at Brixton and 116 days at West Norwood. Although it is clear that visibility varied enormously across London these discrepancies are more likely to be due to differing definitions of what constitutes a fog. The two series are correlated and both show a steep decline starting in 1893 (Fig. 1).

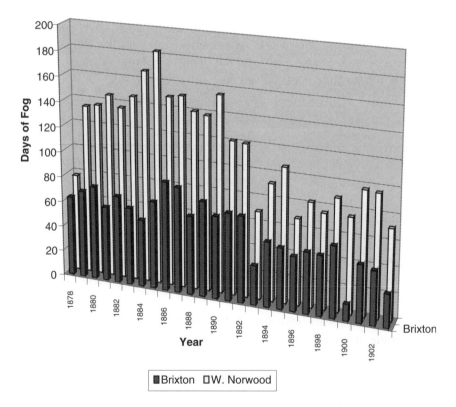

Fig. 1 Days with fog in London at Brixton and West Norwood 1878–1903.

The modern definition of fog used by today's meteorological observers is based on visibility in metres and distinguishes between Dense Fog <40 m; Thick Fog 40–200 m; Fog 200–1000 m and Mist 1000–2000 m. It is not clear what definition of fog was used by the Meteorological Council in Brixton or by Mr Marriot in West Norwood.

During the winter of 1901 and 1902 'The London Fog Inquiry' was carried out by the Meteorological Council funded by a grant of £250 from the London County Council (Shaw 1901). Gaslights were gradually yielding to electric lights and the new owners of the electricity generators were frustrated by unexpected fluctuations in demand due to the changing daylight as the fogs displayed remarkable spatial variation. One house would be in darkness whilst the sun shone upon another across the street. It was estimated that during fogs more than £10 000 a day was being spent on additional lighting in daylight hours. The Fog Inquiry was not set up to research methods of getting rid of the fog but to see if its formation and movements could be forecast up to an hour in advance (Bernstein 1975).

Captain Carpenter was hired to carry out the inquiry and quickly established an observing network of 30 fire stations throughout the County of London. Daily sheets were to be completed and a definition of surface fog was agreed. Light Fog — not sufficient to require artificial light in the daytime; Moderate Fog — when ordinary gas lamps, though visible at 60 yards, are invisible at 440 yards and Thick Fog — in which street gas lamps are invisible at 60 yards or less. This definition led to problems however as several of the fire stations were in narrow streets where there was no clear view. A new more practical definition was then derived (Bernstein 1975:199):

> *Thin Fog or Mist was defined as visibility of objects at 200 yards or more, slightly hindering traffic by rail and river but not by road. Moderately Thick Fog was taken to mean that observers were unable to discern a man by day more than 100 yards away, a house at 200 yards, or a street light by night at 440 yards. Dense Fog meant inability to discern objects across the road by day or lights of street lamps 60 yards distant by night.*

This shows the difficulty of getting observers to agree on the occurrence of fog. The findings of the Inquiry (Carpenter 1903) were broadly as follows. Firstly that the London fogs were locally generated and not imported from the Essex marshes (marshes were a favourite source for weather according to Aristotle), nor did they travel up or down the Thames.

Secondly that there was some evidence that fogs formed in some areas before others and that the fogs moved but no regular pattern of movement was found. Thirdly that in winter light fogs, largely caused by smoke, were permanent in parts of London and that the best visibility in London was no more than 1 1/2 miles.

> *The contamination of the air by smoke has been very forcibly brought to my notice by the ascents of Victoria Tower and of St. Paul's. In the 10 ascents made as yet, none of which were made during fogs, and several of which were made on days of great visibility in the country, the visibility has ranged from 1/2 mile to 1 1/4 miles only. St. Paul's has not yet been seen from Westminster nor Westminster from St. Paul's, although their distance apart is but 1 1/2 miles. (MPMC 1902:114)*

Fourthly if the minimum temperature was above about 42 °F then fogs were unlikely to form. Fifthly if the wind speed was above 13 miles per hour then the fog dispersed. These results were interesting but did not suggest a way of forecasting the fog accurately. The Meteorological Council decided to extend the survey for another winter, at their own expense, but the results did not lead to any prospect of prediction. It is ironic that the first scientific study of the London fog should take place at a time when the fog frequency was declining for the first time in several hundred years.

The London fog was making its mark in other areas, particularly in literature and its description included other features apart from the poor visibility. For example the fog could take on a variety of colours according to the time of day and the thickness of the fog. Ackroyd (2000:432) gives a useful list from a variety of sources:

> *There was a black species, 'simply darkness complete and intense at mid day'; bottle green; a variety as yellow as pea soup, which stopped all the traffic and 'seems to choke you'; 'a rich, lurid brown, like the light of some strange conflagration'; simply grey; 'orange-coloured vapour'; a 'dark chocolate-coloured pall'.*

The Particular became a source of vivid literary description such as Benson (1905)

> *A sudden draught apparently had swept across the sky, and where before the thick black curtain had been opaquely stretched, there came sudden rents and illuminations. Swirls of orange-coloured vapour were momentarily mixed with the black, as if the celestial artist was trying the effects of*

some mixing of colours on his sky palate, and through these gigantic rents there suddenly appeared, like the spars of wrecked vessels, the chimneys of the houses opposite. Then the rents would be patched up again, and the dark chocolate-coloured pall swallowed up the momentary glimpse. But the commotion among the battling vapours grew ever more intense: blackness returned to one quarter, but in another all shades from deepest orange to the pale grey of dawn succeeded one another. (Quoted in Brimblcombe 1988:125)

Brimblecombe (1988:125) suggests that the fine smoke particles in the atmosphere filtered out the blue wavelengths of the sunlight above the fog in such a way that the ground was illuminated by yellow and orange light and occasionally green. At night the fog appeared yellow due to the yellow light of the gas lamps and glare from shop windows.

The fogs were sometimes accompanied by oily, tarry, yellow deposits as noted by Conan Doyle in the Sherlock Holmes story 'The Adventure of the Bruce-Partington Plans' set in November 1895: "We saw the greasy heavy brown swirl still drifting past us and condensing into oily drops upon the window panes." (Doyle 2002:390)

The London fog had became an integral character of Victorian fiction. It became a metaphor that embodied confusion, foreboding and uncertainty about the future. Several stories such as Robert Barr's 'The Doom of London' published in *The Idler* in 1892, predicted that the Metropolis would be wiped out by asphyxiation in an everlasting fog. Ruskin's non-fiction confirmed such gloom and openly declared that the fog was God's punishment. Wheeler (1995:166) points out that Ruskin's apocalyptic view refers to the last book in the Bible 'in which the world is brought to judgement and the glory of God is uncovered in the last days'. Perhaps the greatest novel that hinges on the London Particular is *The Strange Case of Dr Jekyll and Mr Hyde* by Robert Louis Stevenson published in 1886.

Not all writers agreed however that the fog was evil, for example Dziewicki (1902) wrote a chapter in a book on London by 'Famous Writers' entitled 'In praise of London Fog'. To some foreign visitors it was a definite winter tourist attraction and made London seem more immense and unique than ever.

Of course the fog appealed to artists. Turner was the first great artist to paint London fogs and the term 'phantasmagoria' was totally appropriate in relation to his skies (Thornes 1999). The famous image of Maidenhead bridge over the Thames: *Rain, Steam and Speed* (1844) shows how the smoke from the railway engine is blended into the opaque atmosphere.

Whistler, an American artist living in London, also loved the London atmosphere. His *Nocturnes* utilised the night-time fog to give an eerie light and colour to the scenes. Chaleyssin (1995:41) suggests that 'The *Nocturnes* force the viewer to try and enter into the picture, to penetrate the fog'. *Nocturne in Black and Gold: The Falling Rocket* (1875) was criticised by Ruskin in his magazine *Fors Clavigera* (Letter 79:11):

> *I have seen and heard much of cockney impudence before now, but never expected to hear a coxcomb ask two hundred guineas for flinging a pot of paint in the public's face.*

Whistler was outraged and sued Ruskin for libel. The libel action opened on November 25th 1878 and lasted for two days. Ruskin was too ill to attend but witnesses included Rossetti and Edward Burne-Jones. The jury awarded in Whistler's favour the symbolic damages of one farthing but no costs, which soon bankrupted him.

Whistler was a great friend of Monet's and introduced him to the Savoy hotel with its superb accommodation and wonderful views of the Thames (Shanes 1994).

Monet was obsessed with the weather and its changing moods. He has immortalised the London fogs, not only in his London Series but also in his letters (Kendall 1989):

> *... I so love London! But I love it only in the winter. It's nice in the summer with the parks, but nothing like it is in the winter with the fog, for without the fog London wouldn't be a beautiful city... (quoted in Shanes 1994:130)*

> *I love London... I adore London. But what I love more than anything... is the fog. (quoted in Shanes 1994:114)*

> *The Thames was all gold. God it was so beautiful... I began to work in a frenzy, following the sun and it's reflections on the water. (quoted in Kendall 1989:191)*

Monet first visited London during 1870–1871 as an exile from the Franco-Prussian war. During that time he painted three views of the Thames cloaked in winter fog, including The Thames and the Houses of Parliament.[h] These foggy views of the Thames were a precursor of the smoke laden Impression, Sunrise (1873), a view of the Seine at Le Havre,

---

[h]This painting is in the National Gallery, London. To see images of all of Monet's works see Wildenstein (1996).

from which the word Impressionism was derived. Monet was determined to return to London in the 1890s to embark upon a 'London Series'. He had already completed a number of other 'Series' such as The Grainstacks, the Poplars near Giverny, Rouen Cathedral and Mornings on the Seine. Why he chose to go to London and paint bridges, factory chimneys and railway engine smoke after such blissful rural series is open to debate. Monet was resolute that he would make his mark in England and produce a Series that would rank him alongside Turner as one of the greatest landscape painters.[i] Monet also respected the English and London and the political stability that the Houses of Parliament represented at a time of political turmoil in France. In the middle of September 1899, Monet arrived in London on holiday with his wife and her daughter Germaine to visit his son Michel who was in London to improve his English. He took a suite on the sixth floor of the Savoy hotel, on the north bank, with a balcony overlooking the river. Below him to the east was Waterloo Bridge and to the south Charing Cross Bridge. The Houses of Parliament were just a kilometre to the south. From his room he could see the sunrise over Waterloo Bridge and by midday the sun would by fully illuminating the Thames from the south silhouetting Charing Cross Bridge.

During this visit Monet painted exclusively the Charing Cross Bridge with the Houses of Parliament in the background. The winter fog set in during October 1899 and the weather observations recorded at 8 am at Brixton show that there were 14 mornings when fog or mist was recorded (7th–11th and 17th–25th) and on the 17th October Monet wrote that he was 'trying to do a few views of the Thames.' He was entranced with the images before him and although he left London in early November he was determined to return as soon as possible. He went back to France with at least 11 canvases, all of which he was still working on.

He returned to London alone, in the middle of February 1900 and embarked upon his major London campaign, this time from the fifth floor of the Savoy. He began work on 11th February and his working habits are well documented in the many letters he sent home to his wife.[j] He tried to stick to a strict daily routine working on Waterloo Bridge in the early morning as the sun moved around from the southeast. Then moving onto

---

[i] A thorough discussion of the various reasons why Monet undertook the London Series is given by Tucker (1989).

[j] The following quotes are taken from Kendall (1989).

Charing Cross Bridge for the midday and early afternoon sun. Then he would pack up his things and move to a small balcony at St Thomas's hospital opposite the Houses of Parliament to catch the sunset. He would work on a number of canvases each corresponding to a position of the sun in the sky and rapidly progress through them, adding a bit more each day. However as the weather conditions were different each day it was a struggle to maintain such a demanding schedule:

> *Wednesday 14th February 1900*
> *'I had a better day than I expected, I was able to work before and after lunch from my window and at 5, with the sun setting gloriously in the mist, I started work at the hospital.' (188)*
>
> *Monday 26th February 1900*
> *'In the early hours of this morning there was an extraordinary yellow fog; I did an impression of it which I don't think is bad.' (188)*
>
> *Friday 9th March 1900*
> *'... as I had predicted, the sun already sets a long way from the place I'd wanted to paint it in an enormous fireball behind the Houses of Parliament; so there must be no further thought of that.' (189)*
>
> *Sunday 18th March 1900*
> *'The only shortage I have is of canvases, since it's the only way to achieve something, get a picture going for every kind of weather, every colour harmony... I have something like 65 canvases covered with paint...' (189)*
>
> *Monday 19th March 1900*
> *'it became terribly foggy, so much so that we were in total darkness, and I had to have the lights on until half-past ten; then I thought I'd be able to work but I've never seen such changeable conditions and I had over 15 canvases under way, going from one to the other and back again, and it was never quite right; a few unfortunate brushstrokes and in the end I lost my nerve and in a temper I packed everything away in crates ...' (189)*
>
> *Wednesday 28th March 1900*
> *'Just imagine, I'm bringing back ... eighty canvases.' (190)*

Monet returned to France in the first week of April 1900 and didn't return until the end of January 1901 again staying alone on the fifth floor of the Savoy.

> *Sunday 3rd February 1901*
> *'I can't begin to describe a day as wonderful as this. One marvel after another, each lasting less than five minutes, it was enough to drive one mad. No country could be more extraordinary for a painter.' (191)*

Saturday 2nd March 1901
'The weather's terrible... Torrential rain and it's beating so hard on the
windowpanes that I can barely see anything at all. Yesterday I was happy
and full of energy and was looking forward to a good day; yesterday
evening the weather was perfect, but as I've said before, it is not possible to
work on the same paintings two days in succession...' (192)

When Monet returned to Giverny at the beginning of April 1901 he had in total more than 100 canvases of the Thames, of which 95 survive today spread around the art galleries of the world. He was far from satisfied with them, however, and he continued to work on them, from memory, over the next few years. This could have created a dilemma for Monet, the impressionists were only supposed to paint from nature, but the many months immersed in the London atmosphere had given him sufficient experience to finish the pictures in the studio. Just twelve of the 95 surviving pictures are dated at between 1899 and 1901, 57 between 1902 and 1904 and 26 are undated, suggesting that most of those that were dated, were done so when Monet sold them rather than when he painted them. However it is impossible to say how many of the canvases brought back from London were actually finished and how many Monet painted entirely in Giverny.

On 9th May 1904 an exhibition of 37 London pictures opened in Paris entitled *Vues de la Tamis à Londres (1900–1904)*, there were 8 views of Charing Cross Bridge, 18 of Waterloo Bridge and 11 of the Houses of Parliament. The exhibition was an immediate success and the critics duly declared that Monet's atmospheric renditions of London were equal to those of his hero Turner. It was also a commercial success and Monet sold 23 of the pictures to the gallery owner Durand-Ruel who sold many of them on to the United States. Monet had planned an exhibition in London in 1903 but it had failed to materialise — otherwise perhaps some of the pictures, (rather than none), would be hanging in London Galleries today.

We must now ask ourselves how accurate a rendition of London's winter skies 1899–1901 do Monet's paintings provide? Are they accurate impressions or just commercial representations of a popular theme? Monet was asked many times to defend his London and other pictures, especially the range of colours. In January 1905 an article appeared in a French magazine *La Revue Blue* entitled *La Fin de l'impressionnisme* (The End of Impressionism) and subsequent discussions included a criticism of Monet's Rouen Cathedral paintings which suggested that they had been copied from a photograph. One could hardly suggest that the London Series were copies of photographs but Monet was not worried and wrote to Paul Durand-Ruel on 12th February:

> You are quite wrong to worry about what you tell me, indications of nothing but bad feeling and jealousy which leave me quite cold. I know ... only Mr Harrison, whom Sargent commissioned to do a small photo of Parliament for my benefit which I was never able to use. But it is hardly of any significance, and whether my Cathedrals and my London paintings and other canvases are done from life or not, is none of anyone's business and is quite unimportant. I know of so many artists who paint from life and produce nothing but terrible work.[k]

There are two aspects of Monet's paintings of London that we can examine for truth: Firstly, the visibility in the pictures and secondly, the colours. It has already been stated that the London Fog Inquiry in the winter of 1901 and 1902 showed that the winter visibility in London at that time was never more than about 2 km. The visibility depicted in Monet's Charing Cross Bridge pictures can be gauged by estimating the furthest point clearly visible in the pictures which gives a mean visibility of 1127 m (maximum 2000 m, minimum 600 m).[l] (See Fig. 2). These estimates are taken from photographs of the pictures but nevertheless show that Monet's pictures are entirely consistent with contemporary meteorological accounts. He did not paint in dense fogs, as he would not have been able to even see the bridges. The ideal weather was when the sun was breaking through the mist and smoke and the visibility was about 1 km and he could just see the Houses of Parliament in the background. The best visibility of 2000 m is shown in the undated *Charing Cross Bridge, Reflections on the Thames* now owned by the Baltimore Museum of Art, which clearly shows a visibility beyond the old Lambeth suspension bridge which was about 1700 m south.

The colours in the pictures are much more difficult to estimate without access to the originals. The range of colours describing the London fogs has already been given and although Monet may have exaggerated what he saw

---

[k]Kendall (1989) p. 196.

[l]These visibilities have been estimated from the illustrations in Wildenstein (1996). This exercise would have been conducted with the real paintings but they are not accessible in sufficient numbers. The illustrations are good enough to decipher buildings, bridges and the Houses of Parliament etc so the results are unlikely to be significantly affected.

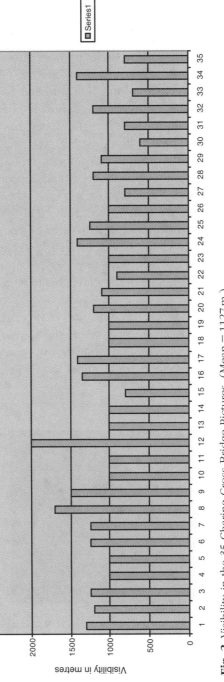

Fig. 2 Visibility in the 35 Charing Cross Bridge Pictures. (Mean = 1127 m.)

his basic colours are true. When he was asked to list the composition of his London Series palette he replied: 'I use flake white, cadmium yellow, vermillion, deep madder, cobalt blue, emerald green and that's all.'[m]

The colours can be divided between the water, the sky and the bridges and buildings. In many of them the sky is directly reflected in the water and the colour of the whole picture is controlled by the colour of the atmosphere. In others the sun is visible, or sunlight bursts through the mist to give islands of colour. Like Turner, Monet has portrayed unique moments of colour and time. Like Constable, whose skies are so typically English, Monet has also portrayed the general typical winter climate of London at that time, which any Londoner would have instantly recognised.

Monet's London Series was painted at a time when the fog frequency in London was declining, having peaked in the mid 1880s. They serve as a constant reminder as to how polluted the city of London was during its glory days as the capital of the world. It was a price that Londoners were prepared to pay and we would have forgotten about the full glory of the London Particular had it not been for a Frenchman who was obsessed with the weather.

Having discussed Figure 3 detail the sites of production and sites of the images themselves it remains to state where the paintings can be seen. Of the 95 paintings in the London Series 41 are in private collections, which are occasionally lent to galleries and exhibitions. Of the remaining 54 there

Fig. 3 Houses of Parliament Sun Breaking through Fog: Image Rights Department, The Art Institute of Chicago.

---

[m]Kendall (1989) p. 196.

are 29 in American galleries and the rest are spread around the galleries of the world (France 7, Japan 4, Canada 3, Switzerland 3, Germany 2, Russia 2, Denmark 1, Egypt 1, Ireland 1 and Wales 1). Not one is housed in an English gallery although currently (in 2004) *Houses of Parliament, Sunset* owned by a Japanese private collector, is on temporary loan at the National Gallery in London. It is as if Londoners do not want to be reminded of their foul polluted past. Of the American galleries The Art Institute of Chicago owns three pictures, one of Charing Cross Bridge, one of Westminster Bridge and one of the Houses of Parliament. It is the only gallery in the world to have a complete sample of the series! The National Gallery of Art in Washington DC has four pictures — three of Westminster Bridge and one of the Houses of Parliament. The only other gallery to have three pictures is the Musée Marmottan in Paris — two of the Charing Cross Bridge and one of the Houses of Parliament. It is ironic that this remarkable London Series, unlike the London Particular, has been exported to the rest of the world.

The deconstruction of Monet's 'London Series' has shown that they do represent an accurate souvenir of the perils of air pollution in London. As such they represent a unique representation of the winter climate of London at that time with a strong subjective element so that we can feel what it was like to be there, to see how the light was diffracted in the smoky atmosphere and to see the golden reflections from the surface of the river. The smokey scenes of that time should remind us how today there are still many pollutants in the atmosphere, even if they are mainly invisible. These can also be dangerous because their concentrations often exceed the air quality standards set for a healthy atmosphere. The air that we breathe is amazingly complex and its composition has evolved along with life over millions of years. We must ensure that its quality is sustained. The glorious colours and mystique of Monet's paintings of the London fog may awaken a nostalgic yearning. However we should really be striving to restore London's atmosphere to the state it had originally, well before the days of the 'London Particular'.

REFERENCES

Ackroyd, P. (2000) *London — The Biography*, London, Vintage Press.
Barr, R. (1892) "The doom of London," *The Idler*, pp. 397–409.
Benson, E. F. (1905) *Images in the Sand*, London, Heinemann.
Bernstein, H. T. (1975) "The mysterious disappearance of Edwardian London Fog," *The London Journal* **1**, 189–206.

Bonacina, L. C. W. (1939) "Landscape meteorology and its reflection in art and literature," *Quarterly Journal of the Royal Meteorological Society* **65**, 485–497.

Brimblecombe, P. (1988) *The Big Smoke, A History of Air Pollution in London since Medieval Times*, London and New York, Methuen.

Brimblecombe, P. (2000) "Aerosols and air pollution in art," *Proceedings of the Symposium on the History of Aerosol Science*, Vienna.

Brimblecombe, P. and Ogden, C. (1977) "Air pollution in art and literature," *Weather* **32**, 285.

Brodie, F. J. (1892) "The prevalence of fog in London during the twenty years 1871–1890," *Quart. J. Roy. Met. Soc.* **18**, 40–45.

Brodie, F. J. (1905) "Decrease of fog in London in recent years," *Quart. J. Roy. Met. Soc.* **31**, 15–28.

Brüggemeier, F.-J. (1994) "A nature fit for industry: the environmental history of the Ruhr Basin 1840–1990," *Environmental History Review* **18**, 35–54.

Carpenter, A. (1903) *London Fog Inquiry*, Report to Meteorological Council, HMSO.

Chaleyssin, P. (1995) *James McNeill Whistler — The Strident Cry of the Butterfly*, Bournemouth, Parkstone Press.

Cosgrove, D. and Thornes, J. E. (1981) "Of truth of clouds: John Ruskin and the moral order in landscape," Chapter 2 in *Humanistic Geography and Literature*. D. C. D. Pocock (ed.) London: Croom Helm and Totowa, pp. 20–46.

Doyle, A. C. (2002) *The Complete Works of Sherlock Holmes*, New York, Gramercy Books.

Dziewicki, M. H. (1902) *In Praise of London Fog*. In *London — as Seen and Described by Famous Writers*, E. Singleton (ed.) New York, Dodd Mead.

Evelyn, J. (1661) *Fumifugium, or the Inconvenience of the Aer and Smoak of London Dissipated*. London: printed by W. Godbid for Gabriel Bedel and Thomas Collins.

Gedzelman, S. D. (1991) "Atmospheric optics in art," *Applied Optics* **30**, 3514–3522.

Hambyln, R. (2001) *The Invention of Clouds*, London, Picador Press.

Howard, L. (1818) *The Climate of London*, **1**, London, W. Phillips.

Janković, V. (2000) *Reading the Skies — A Cultural History of English Weather 1650–1820*, Manchester, Manchester University Press.

Kendall, R. (1989) *Monet by Himself*, Boston, Little, Brown and Company.

Lamb, H. H. (1967) "Britain's changing climate," *Geographical Journal* **33**, 445–466.

Luckin, B. (2002) "Demographic, social and cultural parameters of environmental crisis: the great London smoke fogs in the late 19th and early 20th centuries," in *The Modern Demon: Pollution in Urban and Industrial European Societies*, C. Bernhardt and G. Massard-Guilbaud (eds.), Clermont-Ferrand, Blaise-Pascal University Press.

Mosley, S. (2001) *The Chimney of the World: A History of Smoke Pollution in Victorian and Edwardian Manchester.* Cambridge, White Horse Press.

MPMC (1902) *Minutes and Proceedings of the Meteorological Council 1901–1902*, 17 January 1902, Met Office Library, Exeter.

Neuberger, H. (1970) "Climate in art," *Weather* **25**, 46–56.

Nochlin, L. (1966) *Realism and Tradition in Art 1848–1900*, London, Prentice-Hall.

Pevsner, N. (1955) *The Englishness of English Art*, London, Penguin Books.

Richardson, L. F. (1922) *Weather Prediction by Numerical Process*, Cambridge, Cambridge University Press.

Rose, G. (2000) *Visual Methodologies*, London, Sage Publications.

Rubin, J. H. (1996) "Realism," in *The Dictionary of Art*, **26**, J. Turner (ed.), London, Grove Publishers.

Ruskin, J. (1884) *The storm Cloud of the 19th Century*, London, George Allen.

Schneer, J. (1999) *London 1900*, New Haven and London, Yale University Press.

Shanes, E. (1994) *Impressionist London*, London, Abbeville Press.

Shaw, W. N. (1901) "The London Fog Inquiry," *Nature* **LXIV**, 649–650.

Shelley, P. B. (1819) *Peter Bell the Third*, Part 3, Hell, Verse 1.

Stradling, D. (1999) *Smokestacks and Progressives: Environmentalists, Engineers and Air Quality in America, 1881–1951*, Baltimore, John Hopkins University Press.

Thornes, J. E. (1979) Landscape and Clouds, *Geographical Magazine* LI **7**, 492–499.

Thornes, J. E. (1999) *John Constable's Skies*, Birmingham, University of Birmingham Press.

Thornes, J. E. and McGregor, G. R. (2003) "Cultural climatology," Chapter 8 in *Contemporary Meanings in Physical Geography*, S. T. Trudgill and A. Roy (eds.), London, Arnold, pp. 173–197.

Tucker, P. H. (1989) *Monet in the '90s*, Exhibition Catalogue. Boston, Museum of Fine Arts.

Wheeler, M. (1995) "Environment and apocolypse," in *Ruskin and Environment*. M. Wheeler (ed.), Manchester and New York, University of Manchester Press, pp. 165–186.

Wilde, O. (1889) *The Decay of Lying*. In *The Works of Oscar Wilde*. 1987 Edition, Leicester, Galley Press.

Wildenstein, D. (1996) *Monet I–IV*, Taschen, Wildenstein Institute.

Wilton, A. (1979) *The Life and Work of J. M. W. Turner*, London, London Academy Edition.

Chapter 9

Biodiversity and the Urban Environment: Benefits, Trends and Opportunities

Paul Henderson

INTRODUCTION

Large cities and richness of the natural environment may appear to be incompatible. However if the planning and policies allow it urban areas can be centres of biodiversity with unique attributes which greatly enhance the lives of the inhabitants. The whole sustainability of the city and its surroundings benefit in ways which we are only just beginning to understand.

Biodiversity can be simply defined as the range of living things — plants and animals, together with their myriad habitats — that occupy a locality or region. By comparison with many other large cities in the UK and the world, London is rich in biodiversity, and to an extent it is an example to other cities. This richness stems from its diverse habitats including its railway embankments, its parks, botanic, private and other gardens, its cemeteries, its lakes, rivers and waterways, as well as its tree-rich squares and crescents. Habitat is the essential ingredient for biodiversity to exist, and the habitats in urban environments can be almost countless and sometimes surprising. Life, in some form or another, will attempt to occupy any niche that is available. So, London will be used as a focus for this chapter but much of what is written can be applied to other cities in the UK.

For London, habitat may have seemed to decline significantly over the last 50 or so years as building and other developments in and around the capital have occurred. For example, a quick study of the one-inch Ordnance Survey maps of West London from around 1925 to the present day shows the significant loss of green areas. But those areas that remain, including the most important such as Richmond Park, are substantial and support a significant amount of biodiversity. After a surge of development just pre- and post-war there has followed a period of relative stability in the proportion of London's surface area covered by green spaces and open water compared with other uses. It is estimated that today this is about two-thirds. The 33 Greater London Boroughs cover nearly 158 000 hectares. More than 40% of the total land area is green open space and nearly half of that is considered valuable as wildlife habitat (Anon 2000).

LONDON'S BIODIVERSITY

What do the areas of green space and open water contain? Some examples are:

- More than 1500 plant and more than 300 bird species estimated to have been observed in recent years.
- More than 11 000 hectares of meadows and pastures, containing a significant diversity of species.
- Private gardens, comprising a significant total area and of mixed habitats supporting birds, butterflies and much more besides.
- Sites of Special Scientific Interest (SSSI) (as designated by Nature Conservancy for UK Government) — some in chalk grassland and in woodlands — and Sites of Importance for Nature Conservation — some alongside railways. Also some sites — larger ones — that are Special Areas of Conservation under the EU Habitats Directive.
- Areas important for their birdlife — e.g. parts of the Thames Estuary and Marshes.
- Some rare species such as the stag beetle (found in parks) and the black redstart, especially in the Thames Corridor Wasteland.

Encouraging reports in the Press or elsewhere are not uncommon. Early in 2002, an article in the Evening Standard entitled 'Townies don't need the Countryside' (G. Weightman 2002) reported on the presence of the rare bittern in the relatively new London Wetland Centre in Barnes. The same reporter wrote of how from his own home window in Highbury he enjoyed seeing in just ten minutes a mistle thrush, a flock of long tailed tits, a whitethroat, a cormorant, Canada geese and a blackbird. Others give similar accounts as more areas in the capital develop some opportunities for biodiversity to increase or at least to be sustained at current levels.

These opportunities arise from the actions of many — both through the awareness of more people for the need to respect biodiversity and by the specific actions of authorities and agencies. For example, the Royal Parks Agency (Chapter 11) is taking actions, such as reduced or restricted mowing, to promote sustained habitats for some species. The populations and diversity of species will almost certainly increase as a result.

VALUES

It is not uncommon to ascribe values to biodiversity, in the broad sense, so as to help define its goodness and utility. Here a statement about the values has been developed for the urban environment.

Ethical value. which may include religious and experiential aspects. We have an innate respect for biodiversity, including its origins. Humankind is an essential part. This respect can apply as much to the biodiversity in our cities and towns as to that in the countryside or other open spaces.

Aesthetic value. People appreciate the variety, beauty and changing nature of the biodiversity around them whether it be in parks, in their own gardens or even along the roadside. They can be uplifted by it. They can gain immense pleasure by simply being within it or associated with it.

Informational value. Biodiversity clearly has an educational use. Direct observations and interactions with it lead to an understanding of biological systems and the nature of habitats. In addition, the specific make-up and health of local or regional biodiversity can be of great value in assessing the quality or condition of that local environment, whether it be the air, water, or soil. Biodiversity also provides an orientational value — in space and time. Through it, for example, the seasons have a greater meaning and impact.

Utility and economic value. For the urban environment this will apply mainly to specific activities such as tourism.

Ecological value. Habitats are often relatively complex systems — components depend on and interact with each other. Species are sustained by their habitats. Furthermore, the quality of our environment depends on the interactions of living organisms with the air, water, soil etc. of the many ecosystems within the urban environment and beyond. The impact of these interactions may not always be apparent to us. For example, trees act to remove small particles (including particulate matter having an aerodynamic diameter of less than 10 micrometres or PM_{10}) from the air by incorporating them in the rough surfaces of their leaves. Although these particles may subsequently be washed off by the rain, they are less available to be breathed in by us. City atmospheres may often contain a range of particles that can be adverse to health. The presence of trees in the city undoubtedly improves the quality of the air we breathe.

All these values apply to our urban biodiversity. They are of course more than sufficient for us to take this issue very seriously indeed.

LONDON IS NOT AN ISLAND

Whilst we can take comfort from the fact that urban biodiversity is of great value, unfortunately the news about its sustainability is not all good. The London environment is affected by what can readily come into it — including that which is unwanted. Factors which can impact on urban animal and plant life are introduced through air, water and land.

Species that are new to the country can be introduced through unintended association with trade, particularly involving shipping from more distant countries. If the environmental conditions are appropriate the introduced species can survive and spread. A few introduced species can be particularly invasive and their eradication could be a difficult and long-term problem. The Chinese mitten crab is an example. The original source of this species is probably from China via ship ballast water. Its population is now growing significantly along the Thames catchment from as far as Staines in the west to Tilbury in the east (Figs. 1 and 2). It can cause significant damage to the river banks through its burrowing habit. It also displaces local species and even, through its sheer size, frightens fishermen whom unwittingly haul it up on the end of their lines. Its increase is a matter of concern, especially as it is capable of moving some distances across land. (Clark *et al.* 1998). Species of introduced crayfish are also causing reduction in the long-term viability of some endemic crayfish species.

The butterfly bush, *Buddleia*, which was also introduced from China at the end of the 19th century is a particularly successful invasive species — but one that we seem to have accepted with some pleasure. This is not only because it has helped some butterfly species to survive but also

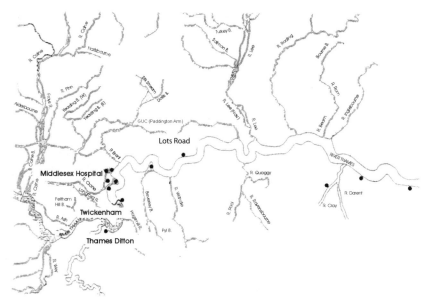

Fig. 1 Distribution of *Eriocheir sinensis* in the Thames catchment from Natural History Museum specimens collected between May 1976 and November 1995.

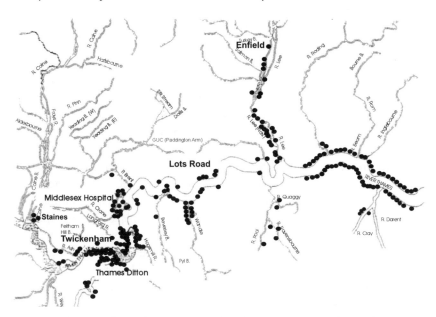

Fig. 2 Distribution of *Eriocheir sinensis* in the Thames catchment from information received between 27 August 1996 and 15 November 1996.

because its common appearance along the rail lines livens up our commuting. On the other hand, the recent introduction of two European vine weevils, possibly through importation of ornamental plants from southern Europe, will not be so appreciated. They have been recorded as causing damage in gardens in several parts of London, especially to laurels, viburnums and some other plants and shrubs with waxy leaves (M. Barclay 2003). They appear to be able to survive an English winter.

The diversity of life in river and stream waters depends, of course, on the quality of the water which is often determined by upstream activities and controls. Indeed the water quality can be assessed simply by the range of fish species within it. The Thames today is reported to contain over 100 species of fish, which indicates that the water quality is reasonable. It certainly compares favourably now and in the past with some other major rivers. For example, the city of Paris, during part of the period when the then 'Monsieur' Chirac was its Mayor (1977–1995), had one of the dirtier rivers in Europe flowing through it. The Seine then contained only three species of fish. Subsequent and significant improvements in pollution control, some introduced by the Mayor himself, have led to an increase in species number of more than ten fold.

Some river and estuary species can be used as indicators of metal and other pollutants in the waters. These kind of studies are particularly relevant because they help provide a link between the nature of the metal in its different chemical compounds with the extent to which it is taken up by the organisms (so-called 'bioavailability'). Certain organisms are now well recognised as useful indicators, or 'biomonitors', of trace metal contaminants in river and estuarine waters. They give an indication of the bioavailability of metals at particular locations over particular periods of time. (See also above under 'values'.)

A recent study by Rainbow *et al.* (2002) of trace metal pollution in the Thames estuary used three biomonitors, a seaweed, a barnacle, and a crustacean, to assess the bioavailability of several metals including zinc, copper, cadmium and lead. The conclusions at this stage are that the bioavailabilities of some of the metals has decreased since the early 1980s but that at certain places (downstream of some sewage work discharges) there are some raised bioavailabilities. Work of this kind is valuable to the understanding of the effects of the reduction or removal of pollution on the return of flora and fauna.

The air quality is, of course, determined by circumstances well beyond the London boundaries as well as by the activities within the capital, especially through the use of the motor car. Plant species that might form part of the urban biodiversity are absent because of poor air quality — the changes in distribution of some organisms, including lichens, can be good indicators or air quality over a suitable period of time (James *et al.* 2002).

It is perhaps too early to say much about the effect of climate change on London biodiversity. Certainly there is evidence of changing distributions of some species in the UK, an example being the extension northwards of the speckled wood butterfly. Milder winters are also adding to the impact of invasive species (such as the vine weevil above) when their survival is now possible.

ACTIONS AND INFORMATION

One of the most important sources of information about the capital's biodiversity is the London Biodiversity Audit produced by the London Biodiversity Partnership (LBP). This partnership was set up in 1996 to identify habitat and species action plans for the capital. The audit is a fascinating read — even for those who possibly have limited interests in biodiversity. It has based its analysis on some twenty habitat and land use types (Table 1) — the categories alone indicate the wonderful diversity in London.

This audit along with other action plans stems from the decisions of the 1992 Earth Summit in Rio de Janeiro, especially those relating to Local Agenda 21 and the Convention on Biological Diversity. The UK took an early lead in producing its own Action Plan (1994) and setting up a steering group to implement it. This group produced its first report in 1995, and it identified *local* biodiversity action plans as the best way forward for the conservation of biodiversity. This is being shown to have been a very wise decision. Biodiversity issues are being placed in the local communities where they belong. Ownership is a key ingredient in developing an understanding of, a caring for, and an enjoyment of, the natural world around us.

At the local level several London Boroughs have published their own action plans and also publicised them on the internet. Many of them give details about the local biodiversity. There is also a wealth of information available on London's biodiversity arising from the enthusiasm and commitment of individuals as well as from organisations such as the London Wildlife Trust and the Field Study Council.

Several organisations are now arranging activities involving biodiversity. Action for Biology in Education, supported by the Chelsea Physic

Table 1 Twenty habitat or land use types for London.

Woodland — all woodland and scrub habitats
Open Landscapes with Ancient/Old Trees — deer parks, old parkland, wood pasture etc.
Hedgerows — all boundary features with trees and shrubs
Acid Grassland — grassland on nutrient-poor, free-draining soils
Chalk Grassland — grassland on chalk
Grasslands, Meadows and Pasture — grassland other than acid grassland, chalk grassland or wet grassland
Heathland — sites where heather occurs naturally
Grazing Marsh and Floodplain Grassland — sites where the habitat depends on periodic wetting or inundation and grazing or cutting
Marshland — all wet terrestrial habitats
Reedbed — sites where common reed is dominant
Rivers and Streams
The Tidal Thames
Canals
Ponds, Lakes and Reservoirs
Private Gardens
Parks, Amenity Grasslands and City Squares — all formally managed amenity open space
Railway Linesides
Churchyards and Cemeteries
Urban Wastelands — semi-natural vegetation on land resulting from previous development or disturbance
Farmland — arable fields and agricultural leys

Garden* has helped organise a 'Backyard biodiversity day' to provide an opportunity for young people to celebrate their own backyard biodiversity and to learn much in the process. The Natural History Museum* (NHM) has made good use of the internet by establishing a survey project on its web pages. This is called 'Walking with Woodlice'* and is now in its third successful year. Its aim is to encourage people to explore their local biodiversity by surveying one particular organism — the woodlice — of which there are probably around 37 species in the UK. The project is aimed at children and teachers but adults also can find it stimulating and fun to participate. Information for teachers and an identification key are provided and the participants feed the local results into the overall survey for the UK. This kind of project has a great deal of value in demonstrating to people something of the nature of the 'diversity' in biodiversity, as well as developing their awareness and some scientific, including data-recording, skills.

Two other initiatives of relevance in which the NHM is also involved are *Flora for Fauna* and the *National Biodiversity Network*. The first aims to encourage people to plant native trees, shrubs and flowers that are local to their area. A Web-based Postcode Plants Database is available that lists and gives images of the native plants and wildlife for any specified postal district. The *National Biodiversity Network** aims to make information about wildlife in the UK (including urban environments) accessible to everyone through the Internet. These and countless other initiatives mean that a general browse through the Web will reveal a wealth of opportunities to learn about the biodiversity of London, or other cities, and to become involved in associated activities.

One key focus today is the overall quality of London's environment. This has stimulated the Mayor of London to produce a biodiversity strategy for London. This is now available on the internet and gives clear actions and mayoral expectations for the years ahead. At this stage we are still mainly defining plans and producing audits. Although there is some criticism of the slow pace, real actions are definitely beginning. Even these could not have happened without the greater awareness of the nature and benefits of biodiversity in the urban environment. Lessons from these first steps are providing a sound basis for establishing a diverse sustainable environment in the long term.

These include:

- Industry and commercial organisations becoming more heavily involved. A few have led the way — Thames Water, for example, is a member of the LBP and is producing its own biodiversity action plan. Other organisations, which do not necessarily have such a direct link to biodiversity, should also become involved because of the intrinsic benefits to all that healthy urban biodiversity gives.
- Biodiversity not becoming too compartmentalised. It must become part of the general planning for the capital.
- Stronger links being made to other environmental aspects especially where they enlighten our understanding of biodiversity. Highlighting local geology, notably London's Chalk, is mentioned in some action plans because of its importance for local flora and soil types. But a broad approach is needed. What about the glacial gravels of Wimbledon

Common, the brickearth of Chiswick, the river gravels of Clapham
Common, and the London Clay of Dulwich?

• Bringing good knowledge of the components of biodiversity into
education and training and commitment and a deeper understanding
of the environment. Educational programmes are needed not only
for school children but also for the public and for borough and city
officials. Species and habitat recognition is a key aspect using
simple identification keys and guides especially for London. Those
working in implementing biodiversity action plans etc., could have
their knowledge of particular taxonomic groups assessed through
the Natural History Museum Identification Qualification (IdQ)
scheme which gives vocational accreditation to people involved in
animal and plant identifications. A scheme, such as that developed
by the Natural History Museum in London, might be initiated to help
deal with London's needs.

• More local authorities should endeavour to include biodiversity as
a central theme in their overall plans for sustainable development,
such as those organised within the common framework of Agenda 21,
established at the Rio Earth Summit in 1992. Many of the initiatives
involve mitigating and adapting to the effects of climate change, which
certainly includes those of the natural environment.

• Consideration of all possible threats to biodiversity, including invasive
species and some burgeoning pest populations. Climate change is
likely to enhance these threats.

A LONDON WILDLIFE GARDEN

Tucked in a corner of the join of Cromwell Road with Queen's Gate in
the South Kensington district, is a small but fascinating area of different
interconnected habitats including woodland, heathland, chalk downland
and fen. It is the Natural History Museum's Wildlife Garden opened in
1995. The different areas and habitats incorporated in the total space of
just about one acre (half hectare) are all now well established and are a
remarkable testimony to the robustness of much of biodiversity as well as
to how conservation of wildlife in the inner parts of our large capital can
be achieved. The location is particularly relevant because Cromwell Road
is one of the busiest and has relatively high levels of associated pollution
products.

A wildlife garden such as this in the heart of an urban environment offers:

- Educational opportunities for local schoolchildren, teachers and others as well as for training in ecological monitoring for the Museum's staff.
- A living laboratory for wide-ranging research on the behaviour in a particularly stressful city environment of the habitats and the permanent or itinerant species. The behaviour can be compared with other parts of London.
- An interesting location for monitoring environmental conditions, determined by the surroundings, notably air quality, and from linking these to the behaviour of the garden and its various components.
- A place for visitors and Museum staff to simply enjoy the richness and atmosphere of habitats with their own balance, in contrast to the much more managed environments of the parks and formal gardens.

The garden has exceeded the expectations of many in the richness of its wildlife. More than 300 species of flowering plant can be seen during the year. Over 50 species of birds have been recorded, as well as more than 300 species of moth and 15 of butterfly. Unfortunately the sparrow, as elsewhere, is now a rare sighting. As to the ponds, chironomids, water boatmen, pond skaters and mayflies were quick to colonise. The ponds also support a large number of dragonflies. (See also Honey *et al.* 1998 and Ware 1999.)

Monitoring of the air quality, which is done in liaison with DEFRA, shows that nitrous oxide and sulphur dioxide levels sometimes exceed the European Union standards — this has led to the loss of certain foliose lichens. Other lichen species, however, are able to withstand the levels and so we are gaining further insights into the suitability of different species to act as indicators of air quality.

REFERENCES

Anon (2000) *Our Green Capital,* Introduction to the London Biodiversity Action Plan.

Barclay, M. (2003) *Otiorhynchus (s. str.) Armadillo* (Rossi 1792) and *Otiorhynchus (s. str.) salicicola* (Heyden 1908) (Curculionidae: Entiminae: Otiorhynchini), two European vine weevils established in the UK, *The Coleopterist* **12**, 2.

Clark, P. F., Rainbow, P. S., Robbins, R. S., Smith, B., Yeomans, W. E., Thomas, M. and Dobson, G. (1998) The alien Chinese mitten crab, *Eriocheir Sinensis* (Crustacea: Decapoda: Brachyura), in the Thames catchment. *J. Mar. Biol. Ass. UK* **78**, 1215–1221.

Honey, M. R., Leigh, C. and Brooks, S. J. (1998) The fauna and flora of the newly created Wildlife Garden in the grounds of the Natural History Museum, London. *The London Naturalist* **77**, 17–47.

James, P. W., Purvis, O. W. and Davies, L. (2002) Epiphytic lichens in London. *Bulletin of the British Lichen Society* **90**, 1–3.

Rainbow, P. S., Smith, B. D. and Lau, S. S. S. (2002) Biomonitoring of trace metal availabilities in the Thames estuary using a suite of littoral biomonitors. *J. Mar. Biol. Ass. UK* **82**, 793–799.

Ware, C. (1999) A survey of vascular plants in the Wildlife Garden of The Natural History Museum. *The London Naturalist* **78**, 35–64.

Weightman, G. (2002) Townies don't need the countryside. Evening Standard.

WEB PAGE ADDRESSES

Chelsea Physic Garden: http://www.chelseaphysicgarden.co.uk/
Convention on Biological Diversity: http://www.biodiv.org/
Flora for Fauna: http://www.fauna-flora.org/
London Biodiversity Action Plan: http://www.lbp.org.uk/ogc/ogc1.htm
National Biodiversity Network: http://www.nbn.org.uk/
NHM Wildlife Garden: http://www.nhm.ac.uk/museum/garden/index.html
Walking with Woodlice: http://www.nhm.ac.uk/interactive/woodlice/

Chapter 10

London's Water Supplies

Roger S Wotton and Helen Evans

WATER SUPPLY IN A SUSTAINABLE CITY

Human settlements are dependent on supplies of fresh water. This was highlighted at the 2002 World Summit in Johannesburg (WSSD) where provision of clean drinking water for more of the World's population was recognised as a priority issue.

In London we turn on a tap whenever we need water for domestic or industrial purposes. It has been the policy in London, as in cities elsewhere in the World, that all water coming from the domestic mains supply is fit for drinking even though only a small percentage (less than 5%) is used for this purpose. Where does this water come from, and how is the huge demand met? Water was pumped from the Thames in 1582 (Sharp, undated), and London's first pipes connected homes to "mains water" in 1619, but what is the mains water system today, and how is it controlled? What of the future, as the population of Greater London expands and individuals use more water, especially for washing and cleaning?

Water for domestic supply originates from rivers and from boreholes driven deep into water-bearing chalk far below ground level and it undergoes careful purification treatment before it is distributed in a form fit for human consumption. Some of the water in the rivers has been discharged from sewage treatment works upstream and this represents one of the most sustainable aspects of urban civilisation, as water is used safely by several population centres along rivers. The rivers also receive runoff from farms and industrial areas and these contribute potential pollutants.

Treated water is passed to the Thames Water Ring Main. This £250 million project, completed in 1994, provides an 80 km pipe network circling some 40 m under London. The Ring Main has a diameter of 2.5 m and its construction was a major feat of civil engineering as the geological formations beneath London are complex and powerful hydraulic forces are generated by the huge volumes of water carried by the Main. From the Ring Main other pipes carry water in a shallower distribution network that eventually ends in the small diameter pipes that feed into houses and other premises.

WATER FROM RIVERS AND BOREHOLES

Most of London's water is supplied from the River Thames and from the River Lee that enters the Thames opposite the Millennium Dome site. The quality of the water from the Thames has varied dramatically over the years. As London's population grew, so did the amount of raw sewage flowing into the river, culminating in the "Great Stink" of 1858 (see Chapter 18). By this time, some wealthy Londoners had their water brought to them from outside London, while citizens from poor communities often drank from wells receiving drainage water from cess pits and graveyards.

Today, the River Thames is the world's cleanest metropolitan river, with 120 species of resident or migratory fish. This is largely the result of improved wastewater treatment and the control of effluent discharges. However, borehole water remains generally of better quality than river water, because of the filtering action of the ground as rain water drains to the aquifer. However, increasing urban development in their catchment area threatens the quality of the water entering these underground aquifers.

Borehole (ground) waters require some purification but much less than the treatment required for river water. Before river water passes to domestic supplies it must be cleaned of inorganic and organic particles and also from dissolved matter. Among the organic particles are bacteria and other pathogenic organisms that cause infections. Dissolved matter includes pesticides, pesticide residues and other harmful chemicals.

River water first passes to reservoirs and, after a period of residence, to Water Treatment Works.

LONDON'S RESERVOIRS

Location and Maintenance of Reservoirs

London's reservoirs store about 30 million m^3 of water and are found in two main regions: in the valley of the River Lea and in West London, near Ashford and Hampton. Reservoirs in the Lea valley include the William Girling, Banbury, Lockwood, High Maynard, Low Maynard, Walthamstow 1, 2, 3, 4 and 5, East Warwick, West Warwick — all supplied from either the Lea or the Thames/Lee tunnel. In addition, the King George V reservoir receives water from the Lee and from the New River — a 400-year-old aqueduct transferring water from the Lea (and from various boreholes) to Hornsey and Stoke Newington East reservoirs. Reservoirs to the west of London, and very familiar to those flying from Heathrow Airport, are the Queen Elizabeth II, Queen Mary, King George VI, Queen Mother, Staines North, Staines South, Knight, Bessborough, Wraysbury and Island Barn Reservoirs. They are conspicuous features and have a large surface area (of about 12 km^2). Unlike some other major cities, such as New York, Los Angeles and Rome, many of London's reservoirs lie within the boundaries of the metropolitan area.

Reservoirs vary greatly in age (1820 to 1970) and require routine maintenance to ensure their safety and good condition. The Reservoirs Act of 1975 requires any reservoir that is above the lowest natural surrounding ground level, and having a capacity of 25000 m^3, to be inspected by an Independent Engineer appointed by the Home Secretary at least every 10 years. In addition, a Thames Water Engineer inspects reservoirs approximately every 6 months, with inspections by other qualified personnel carried out at times ranging from daily to once a week. Reservoirs are not just for water storage, because this is where the process of water treatment begins.

Biological Processes within Reservoirs

Within reservoirs, particles carried by the current of water in the river begin to settle out. However, nutrients and small particles remain in suspension but biological processes within the reservoir steadily reduce their abundance. During the summer months it is usual for large water bodies to become stratified thermally, with warm water overlying cooler water. Warm water is less dense than cooler water so the water column does not mix. With little mixing in the water column the products of decomposition are trapped in deeper water and low oxygen tension occurs when micro-organisms use up available oxygen. This results in water of very poor quality that would be almost untreatable for drinking water supplies. However active mixing of the water causes vertical circulation so as to prevent this problem.

The first area of water processing in reservoirs is at the water surface. All water bodies contain hydrophobic matter that accumulates at the water-air interface and a community of organisms attaches to the surface film. As solar radiation passes through this layer, some photolytic breakdown of organic matter occurs, especially that resulting from the effects of the ultra-violet component in light. This releases labile organic matter that is utilised by bacteria and respired as carbon dioxide that escapes to the atmosphere. Some volatile organic materials also evaporate from the water surface. Light passing into the water is used by algae in photosynthesis, the conversion of carbon dioxide and water into carbohydrates. These, with nutrients taken up from the water column, are converted into chemicals required for growth and metabolism. Algae are thus further diminishing the availability of nutrients within the reservoir water but this is not a constant process as light and nutrients are not in constant supply.

Algae growing in the reservoirs are food for planktonic invertebrates and these plankton are food for other invertebrates and for fish. Animals inhabiting the muddy substratum consume organic matter that settles here and it is also broken down by colonising micro-organisms. All the organisms respire, converting carbon to carbon dioxide gas that escapes to the water and, eventually, to the atmosphere if not taken up during photosynthesis. The net result is that water stored in the reservoirs is cleaner than that entering from the river and this marks the first stage in purification of drinking water. From reservoirs the water passes by pipe or aqueduct to a Water Treatment Works.

DEVELOPMENTS IN WATER TREATMENT

Coarse screens are used to remove branches and any other large objects from the water prior to arrival at a Water Treatment Works (WTWs). Once screened, the water then undergoes several stages of treatment consisting of both physico-chemical and biological processes.

The primary treatment of source water from all surface water drainage involves particle removal by prefilters (also referred to as rapid gravity filters) and, usually, the oxidation of organic molecules by ozonation.

Ozone is a vigorous oxidising agent and breaks up harmful organic molecules, including those of human origin e.g. chemicals from birth control pills and pesticides from gardens or horticulture. Ozonation has been incorporated into the COCODAFF™ (Counter Current Dissolved Air Flotation and Filtration) process at Walton WTW, a high technology solution to water treatment that removes particles and combines organic residues and natural organic matter into flocs that can be collected readily. After treatment with ozone, water is passed either through Granular Activated Carbon (GAC) or passed to Slow Sand Filters (SSFs), some of which contain a layer of GAC (SSF Sandwich™ beds). The reason why carbon is so important is that its abundant adsorption sites allow trapping of any residual harmful chemicals, primarily organics. Carbon also removes the organic molecules that can cause musty tastes and even odour problems once water is poured. Over time, adsorption sites on the carbon become saturated and must be re-activated. Carbon from SSF Sandwich beds must first be separated from the sand and this is carried out on site using a specially designed density separator. The carbon is then delivered to Kempton Park WTW where it is thermally re-activated, a process that removes adsorbed organic compounds. The carbon is then washed and tested for adsorptive capacity before being returned to site.

Research continues on methods to improve the performance of the carbon, the main current development being the introduction of an acid washing stage to remove adsorbed inorganic molecules, mainly calcium. GAC technology certainly brings great benefits to modern water treatment.

"FINISHING" WATER USING SLOW SAND FILTERS (SSFs) AND CHLORINATION

Processes within Slow Sand Filter Beds

Slow sand filters are used to treat water of widely varying quality. They provide a low technology solution for cleaning drinking water of impurities and they are very effective in the removal of pathogens. Each filter consists of a large concrete tank that is layered from the base with porous bricks, then cobbles, a layer of sand, a thin layer of GAC and a final thick layer of sand on the surface of the substratum (some SSFs omit the GAC layer). Water for treatment passes into the concrete tank and flows constantly, the inlet being close to the water surface. Water then drains by gravity through the sand, GAC, sand, cobbles and bricks and then passes through underdrains to the water supply system.

For a newly-built slow sand filter bed to work effectively it must first be "run to waste". This ensures the development of a microbial biofilm over the surface of the sand grains. The micro-organisms, and the exopolymers they secrete, are vital in removing organic matter and a community of organisms that graze the biofilm maintains the microbial community in an active state (Wotton 2002). Other organisms trap micro-organisms and organic materials passing through pores between the sand grains.

Organic matter accumulates at the surface of the filter bed and becomes mixed with the top few millimetres of sand to form a "schmutzdecke" or dirty layer. This is a key feature of the efficient slow sand filter as it is here that most organic impurities are removed. Organic matter and its associated micro-organisms provide food for a range of animals that live here including very large numbers of oligochaete worms and larvae of non-biting midges (Diptera: Chironomidae). These animals alter the composition of the organic matter and probably serve an important role in maintaining flow by keeping pores open. They do this by packaging biofilm and organic layers into faecal pellets and, in midge larvae, by secreting protective tubes of silk, this material being one of the most adsorptive substances known (Wotton 2002).

Some slow sand filters also have abundant growths of algae within the water column and, especially, over the substratum. SSFs are efficient because water is constantly passing through the physical and biological filter, carrying nutrients and oxygen with it to maintain a healthy biological community. This continues as long as pores remain unblocked. A minimum flow is required to prevent the development of anaerobic conditions that adversely affect water taste and quality. On occasions, when the amount of accumulated organic matter is so great that the water no longer passes through the bed efficiently, a bed is drained to allow the surface of the sand to be skimmed away (Fig. 1). Subsequently, the filter bed is then re-filled with water and continues to function, since the microbial community within the bulk of the sand remains intact. These relatively low technology systems (reliant as we have seen on complex biological and chemical processes) can be used effectively in many parts of the world where more costly technologies cannot be afforded.

Recently, experiments have been made by Thames Water in completely covering a slow sand filter bed in order to reduce greatly the light intensity. At the same time it is receiving good quality water, this leads to

Fig. 1 Cleaning the surface of a slow sand filter 50 years ago by manual labour and today by modern plant (adjacent to the bed being cleaned is a slow sand filter in operation).

a considerable reduction of algae in the water and over the sand, and little accumulation of organic matter in the form of a schmutzdecke. Indeed, the surface of the sand in covered beds appears to be clean. However, the biologically active microbial community is still present, as are large numbers of worms and other invertebrates. So covering neither affects the biological nature of the treatment process nor compromises treated water quality. One key difference in the animal community is the absence of insects because covers prevent adult insects from laying eggs into the water.

Chlorination

The final stage of processing involves chlorination of water. Chlorine not only kills micro-organisms in water about to leave the WTW but its presence within the distribution network ensures that no further re-colonisation occurs. The very low levels of chlorination used are difficult to detect by the consumer.

CONCLUSIONS AND FUTURE DEVELOPMENTS

London still has a plentiful supply of fresh water although this is being challenged as the city grows. Water purification processes make use of new technology, but they will continue to be heavily dependent on the natural biological systems that occur in reservoirs, in flocculation and in sand filters. These have been shown to be highly effective, even removing persistent trace chemicals. Much of the engineering in Water Treatment Works, making use of the latest research, is designed to harness and optimise these processes.

As a result of all water treatment processes, drinking water quality in London is maintained at a very high standard. Our ability to measure contaminants has improved over recent years, resulting in greater awareness of the variety of contaminants and their concentrations. Tighter regulations to protect public health have followed. Contaminants now monitored range from heavy metals to pesticides to pathogens. In total, 57 of these are monitored routinely. However, certain complex contaminants, such as pesticides, require the analysis of more than one substance, so tests are carried out on a total of 90 different substances. Results show a very high percentage of tests complying with required standards (Fig. 2).

In rationalising its use of space for reservoirs and WTWs, Thames Water has been able to release land for building and for nature reserves. For example, the Barn Elms site has been converted into a wildfowl refuge that attracts many visitors, both bird and human. The focus remains on providing good quality drinking water and in anticipating change. As Dr Tony Rachwal (Research and Development Director, Thames Water) emphasised at the LEAF conference, efforts will continue to promote the efficient use of water. Drinking water is used widely for bathing, cleaning cars, watering gardens, etc. and we must encourage its more sustainable use.

The future needs of London for drinking water present many challenges but investment in research on treatment and supply technologies

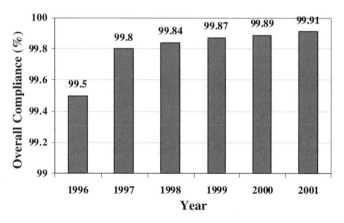

Fig. 2 The percentage of tests on London's drinking water complying with Drinking Water Standards (Ref: Thames Water 2002).

will ensure that London continues to be supplied with safe drinking water. The provision of drinking water to the metropolitan area is a real success story and will continue to be so.

ACKNOWLEDGEMENTS

We would like to thank John Sexton (MD of Thames Water Utilities Limited) for permission to publish this paper and Julian Hunt for his editorial help. The views expressed are solely those of the authors.

REFERENCES

Sharp, J. J. (undated) Water, Sanitation and Health — A review. Website at: http://www.engr.mun.ca/~jsharp/6101/6101.html.

Thames Water (2002) Drinking Water Quality 2001, Environment & Quality, June 2002, Thames Water Utilities Ltd., Clearwater Court, Vastern Road, Reading, Berkshire RG1 8DB.

Wotton, R. S. (2002) Water purification using sand. *Hydrobiologia* **469**, 193–201.

Chapter 11

The Royal Parks and their Role in a Sustainable City

William Weston

Over the summer of 2002 as part of my holiday, I walked through Wainwright coast to coast and walked from St. Bees in the Lake District to Robin Hood's Bay on the Yorkshire Coast — a real and direct engagement with the natural world at several different levels. Travelling on foot — a natural experience in itself — I experienced the wind and rain, saw plenty of birds and other wildlife, and met other people engaged with the countryside. I came back to work with a cleared mind, and perhaps a little fitter (although the Yorkshire breakfasts probably undid much of the good work!).

What opportunities do Londoners have of experiencing even a little of this during an average week; often we are crushed on overcrowded buses and the underground or bumping into people along congested pavements.

The parks in London provide a real opportunity for relief from the busy hubbub. As managers of The Royal Parks — which comprise of 5000 acres of green, open space, gardens, lakes and rivers, right in the heart of London — we have a unique responsibility to Londoners. The challenge we face is enriching and guarding these precious resources against the increasing pressures on the natural environment.

There are many elements that combine to influence the urban city dweller and his or her state of well being. We need to consider health, water, architecture, transport, open spaces, sustainability, noise, air quality, land use, and bio-diversity. Parks have a role to play in each and every one of these topics. Green spaces bring sunshade and temperature reduction, wind and noise reduction, security and pollution absorption. They are also valued for the integration of social and physical activity. As one of the major providers of open space in London the Royal Parks has wide and far reaching responsibilities. This has affected our purpose. As an organisation we have moved from being a management body for a rather special estate to becoming one with a dynamic cultural role at the heart of London life and important relationships with its communities. We are working to explain to the public this new role, in order that they can better understand and enjoy the Parks in the changing environment of London.

ENVIRONMENTAL AWARENESS

The Royal Parks recently commissioned a study of Londoner's attitudes to their open spaces, which showed that they largely take them for granted. Many people had the perception that London's parks, just like hills and rivers, have always existed. This is not true. The parks exist only because of past historical decisions by monarchs and a number of visionaries, who set out to invest for future generations. The parks were created for Londoner's to enjoy, and have required continual management and upkeep to protect them.

How do we heighten awareness of the changing role of open spaces in the urban environment? Charles Kennedy M.P. in a speech to the UK Sustainability Development UK Conference in 2002 answered this. He said:

> 'Environmental awareness has to start at the level of the individual and the community. There's far more that can be done to encourage people to be more conscious of the consequences of their actions for future generations.'

What better circumstance is there for individuals and communities to come together to enjoy environmental awareness than in a public park? As the 'trustees' of London's green, open spaces, the Royal Parks plays a key role in encouraging these practices through working directly with people of all ages and from all walks of life.

If we need any further evidence of the need to heighten environmental awareness, we need only look at research into children's knowledge and understanding of the world they live in. In March 2002 research by experts at Cambridge University into whether urbanisation is reducing children's knowledge of nature in favour of human artefacts was published (based on the 'Biophilia Hypothesis' that humans have an innate desire to catalogue, understand, and spend time with other life forms). Primary school children were surveyed to assess their knowledge of natural history against that of artefacts, in this case Pokemon cards. The survey demonstrated that young children have considerable capacity for learning about creatures natural or artificial, but that they were apparently more able to learn about Pokemon than wildlife. Children about to enter secondary school could name less than 50% of common wildlife species, and this was at the level of identifying them in general terms such as — rabbit, daffodil, mice, beetle or badger. By contrast the Pokemon characters did well!

In the autumn of 2000 there were further survey findings that similarly demonstrated a lack of countryside knowledge among children, such as which tree produced acorns.

Where can children who live in London's inner city learn about nature? The answer is to one of its great parks.

THE IMPORTANCE OF NATURE

These kinds of results beg the question: 'how much do the parents know and think about the natural world, and how much do they care?' When it comes to the grown-ups we get more mixed messages. Since supermarkets now sell the same range of fruit and vegetables from around the world throughout the year, many people have lost any sense of seasonality in terms of what they eat. And our experience of transport is a long way from sharing country lanes with cows, as was the case in London up until 100 years ago. Now we rarely acknowledge nature, barely looking out of the windows of crowded trains and buses, or sharing congested roads and pavements. Do these pressures make Londoner's consider their own everyday environmental choices. Are we even thinking more widely about the environmental choices of our country and the whole world, perhaps stimulated by the World Summit on Sustainable Development held in 2002 in Johannesburg? Alas, despite its importance (see Chapter 3), this event would have passed most Londoner's by.

In the UK workers now tend to work very long hours, which contrasts with gentler, earlier times when there was time for recreation with the kids, such as playing ball games in the open air. This is an important feature of everyday life, with good physical and emotional health being helped enormously by both exercise and being in a green, open space. Indeed, how does our loss of contact with the natural world affect our health? It is thought that the increase in the number of people suffering from allergic reactions might be something to do with our 'too clean' society. Only recently Californian scientists set out to prove that children who had animals in the home had a greater allergic reaction as adults than those who didn't. In fact they proved the exact opposite. Could it be that direct, physical contact with animals, trees, plants, soil and water, have a greater influence on our overall health than we might expect?

When city dwellers lose touch with the natural world the well being of the whole population suffers. This starts with us not being able to understand, enjoy, or nurture the complex and wonderful habitats created by nature. The misunderstandings and clashes of interest in the foot and mouth epidemic and its aftermath provide a sombre lesson for everyone, illustrating a vital need for us to respect our natural environment.

On the other hand there are also signs of a desire for a different and 'greener' life experience by urban dwellers.

Many of us continue to move further out of the city or, if we can afford to, take second homes in the country to gain an improvement in our 'quality of life'. Arguably this leads to damaging effects on countryside and city alike. Within London, houses with good gardens sell at a premium and, of course, so does property near a park. And both farmers' markets and organic goods are gaining in popularity. When families were asked a year or so ago what aspect of their local community was most important in contributing a real benefit to their way of life, the vast majority singled out accessible green, open space. So, maybe, besides those of us who have a deep concern for environmental issues, there is a propensity among Londoner's to engage with the natural world in some form. Perhaps what is needed is a more direct approach by park managers?

DEVELOPMENTS IN THE ROYAL PARKS

Together with the managers of other open spaces within London, the Royal Parks encourages people to engage physically with the green, open spaces we offer. Altogether we encompass eight parks, each with different characteristics for its visitors to enjoy. St. James's Park is an historic landscape, situated by the ceremonial heart to the nation. In fact with over 10 million visitors a year it is one of the busiest parks in the world. Green Park is the only 'simple' area of grass and trees in central London. Hyde Park and Kensington Gardens, which between them provide 48 acres (see Fig. 1) of water bodies (the Serpentine, Long Water and Round Pond), historic trees and avenues, natural habitats and grassland areas for a variety of animal and plant species, gardens, opportunities for children's play, and informal recreation. Regent's Park, a garden park set against Nash's architectural vision,

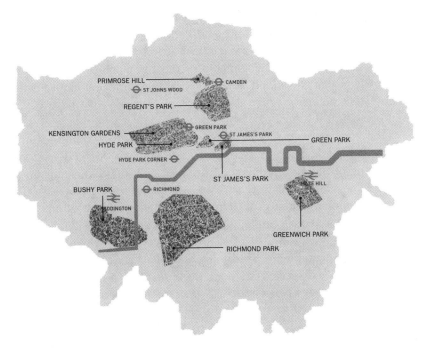

Fig. 1 Locations of London's Royal Parks.

with, additionally, a large area available for formal and informal sport. Greenwich, a historic formal landscape with ancient veteran trees and deer. Outside London's urban centre, the two ancient deer parks of Richmond (a site of special scientific interest) and Bushy Park (the deer park for Hampton Court Palace) provide some of the best opportunities for Londoners to have a Royal countryside close to central London. With acid grasslands in Richmond and watercourses and rivers in Bushy, these parks provide a huge range of resources and opportunities for people to engage thoughtfully in the same landscapes that our earlier monarchs enjoyed. Clearly, as park managers, we have important tasks of conservation and protection, but perhaps our wider role in delivering a practical and informed message to visitors and prospective visitors is even more important.

This involves the Royal Parks developing programmes of education and participation in relation to all the natural environment and recreational facilities that we offer. To date, small educational developments have been possible through local voluntary initiatives. But with the recent appointment of a Head of Education within the Royal Parks the prospect of significant enhancement of education within our core programmes will begin.

In terms of **biodiversity** we actively support the London Biodiversity Partnership. Lowland Wood Pasture and Parkland is the type of habitat in the parks that features in the key national action plan. We seek to attain targets for protection and development of the communities of plants and animals. Given that Richmond Park, for example, contains more ancient trees (by strict definition) than say France or Germany, we have a very special ecological resource to nurture.

We have an active policy in terms of managing grass-cutting regimes. Long, medium, and short-grassed areas are now encouraged in zones around many of the less frequented areas. An inch or so variation in the cutting regime can make a remarkable difference to biodiversity. The wrong management regimes can harm opportunities for animal and plant species to develop. For example, a contractor in an area outside the Royal Parks inadvertently cut an area of long grass very short while testing out new grass cutting equipment. This unintentional act seriously damaged the habitat of a protected species of spider. Co-ordination, communication and active management by parks staff can play an important part in habitat protection.

Knowing what goes on within our park and the habitats they support is vital. Because of sparse resources this requires community involvement. For example, by involving local organisations and individual enthusiasts in data collection we have increased our understanding of the parks' natural environment. This also built community relationships and encouraged the wider public to come into London's parks.

PARK MANAGEMENT

There is not enough detailed research on the overall impact of those human activities that are detrimental to the natural world within our parks. This prevents these activities from being curtailed. For example, we are encouraging the local community around Richmond Park to stop driving through the park (for example, on the commuter run into town) and thus reduce the overall environmental impact made by road traffic in terms of visual intrusion, interference, noise, and exhaust pollution (Department of Transport statistics show that road traffic on average is increasing in the London area). Traffic clearly has a negative impact on the flora and fauna, but can we be specific and quantify its effects? Whilst a good case exists from past research, more detailed and definitive studies are required.

Indeed, most of us are probably unaware of the extraordinary range of wildlife that exists in London's parks, be it nesting skylarks in Richmond, owls in Hyde Park or even the odd kingfisher on its water bodies. In partnership with the Royal Society for the Protection of Birds (RSPB) we will be launching 'Wildlife for All' aimed at putting Londoners in touch with the natural world on their doorstep. This will be developed through a range of new activities run by parks staff working with our Head of Education and Wildlife Officers activities (for example, a new mobile viewing and information point in some parks).

To support and broaden this effort we have recently appointed the first Community Ecologist to the Royal Parks team.

The parks provide an opportunity, gradually being developed, to demonstrate **sustainability** in action. Each park can provide a practical model which can be part of our agenda for community engagement. Recycling green waste and its management is now part of everyday park life and could become a partnership opportunity for other organisations and local authorities around London. Almost all our fleet of vehicles operate on Liquified Petroleum Gas (LPG) and new park buildings such as the Cake House restaurant in St. James's Park and a sports pavilion in Regents Park have been designed both with sustainable and environmental objectives in mind. The three outer parks (Richmond, Bushy and Greenwich) now have official (International Organisation Standardisation (ISO) 140001) accreditation for their environmental management and we are working towards accreditation for all parks by next year.

Developing and fostering the use of pedestrian and cycle routes around the parks is an effective way of encouraging the public to learn about use of parks, even if different users have to watch out for each other. Often the public are unaware that taking a route through a green space is as quick as taking the tube. We are working with Central London Partnership on this agenda, and continually trying to enhance the enjoyment and experience of walking in London.

Our resources, facilities, and public role, provides the Royal Parks with a real opportunity to demonstrate new technology in practice. We are examining the possibility of creating a renewable energy project at Bushy Park, utilising mini hydro-electric generators, solar and wind power, green waste matter, and a variety of other ideas. Not only could this project deliver practical benefits to Bushy Park itself, but it could also provide a working demonstration of different technologies for a range of purposes from commercial sales to a structured educational programme.

A HOLISTIC ROLE FOR THE PARKS?

With doctors and the government calling for the British to adopt more healthy life-styles, parks can play a vital role here as well. In addition to the health benefits of walking and cycling, parks provide the space and opportunity for sport and recreation, especially since school playing fields are disappearing or not being maintained. The range of organised sports activities in the Parks is astounding, from softball to archery, from golf to rugby, and from tennis to football.

But perhaps it is the casual use of the space for informal games which is most important to the Londoner. Indeed, parks provide an opportunity for social interaction of all kinds, when such opportunities are becoming less frequent. Be it office workers decanting into a park during their lunch hour or families exercising and interacting. It's extraordinary to witness Regent's Park on a sunny summer evening, even mid-week, and see the many thousands of people using the park for all kinds of activity. Ball games, a family picnic, an opportunity to sit under a full sized tree. All offer the opportunity to gently engage with other people.

The natural world creates a context for harmony and relaxation, which is intrinsically good for our health. But parks must engage with the public and actively promote themselves. It is known that continued gentle exercise in the elderly maintains mobility through into extreme old age, as anyone visiting a park in Beijing will confirm. In an ageing population the quality of life for the elderly is of increasing importance. Indeed, walking in general is in decline. For example, over 23% of children nationally are now being driven to school. Parks provide the opportunity to walk (and cycling or rollerblading), whether it is through the use of a newly discovered route to work or school, to maintain mobility, or simply for enjoyment.

There are two important walking routes through Central London which use the Royal Parks. From Waterloo to Kensington Palace or Queensway (through St. James' Park, Green Park and Hyde Park) and from Camden to Oxford Circus (through Primrose Hill and Regent's Park). With the new Hungerford Bridge over the Thames, the former route is entirely within the parks and other green public spaces, with only a few road crossing points. The existence of walking routes such as these and the accompanying benefits they offer need to be properly publicised and their use encouraged, for example, through promoting an *overground* map of London.

Often we are aware of the sense of relief when we enter a park, for example, when leaving a busy road outside. The Royal Parks' overarching objectives refer to the 'quiet enjoyment' of the parks. There is something spiritual about this 'communing' with nature. A space that delivers tranquillity is like a work of art! A composition has to be created. Social objectives have to be integrated. There is more to our parks than planting a few trees and some grass, and then hoping for the desired effect.

Stress in our urbanised lives is pervasive, affecting our health to such degrees that the drugs most commonly taken in our society are sedatives such as valium. The daily stress experienced by everyone erodes stamina, a sense of well being, and health generally. Studies of the impact of green spaces on stress have shown that health benefits begin 3 to 5 minutes after entering the space. They include lowered heart rate and blood pressure, decreased tension, and reduced fear and aggression. These effects are sustained after the person has departed.

Britain is renowned world-wide for its variable weather! Only in the parks can the changing seasons and the elements be fully appreciated, rain or shine. Green spaces moderate the local climate. The heat during summer in London can at times be oppressive. The high percentage of green spaces in London benefits the local micro-climate considerably, with typical temperatures 1 to 2 degrees Celsius cooler in the parks. Extensive planting is important and park managers have a role to play in encouraging the greening of London. This is especially important in the immediate vicinity of the park boundaries, where it is possible to create a pleasurable sense of anticipation of entering a green space.

THE ROYAL PARK'S FUTURE AND ROLE

As park managers we need to be pro-active. In some ways the term 'manager' undervalues our role as 'green protagonists'. In many different tasks of park management level we engage individuals and groups day to day in the whole agenda through encouraging participation in a wide range of activities. Our community and education agenda needs to reach those who are less inclined or less able to take advantage of our open spaces. Recruiting volunteers to work in education centres created in each park can help us achieve this goal. We already have successful centres working with primary school children established at the Lookout in Hyde Park, at Holly Lodge in Richm ond Park and at the Stockyard in Bushy Park.

Communicating our message to the public requires imagination and work. Perhaps we need to make 'virtual parks' available on the internet to reach and inform younger generations of the opportunities our parks provide (see also Chapter 15). The skills we have in interpretation and education can equally be applied to the park visitor or green tourism in general.

I believe that increasingly we should be seen as community trainers. This requires us to lead by example. Certainly much is needed to be done to develop our capabilities and raise the esteem of park staff through investment in training and a better salary progression, parks staff being notoriously underpaid! This should encourage skilled individuals to consider green space management as a worthwhile career.

We have a critical defensive role in battling against threats of erosion and encroachment to our green spaces in London. With land at a premium in London, this often involves using planning procedures to question and prevent the (further) development of buildings which might have a negative impact on our experience of our parks. It is not just very tall buildings that impinge on our parks. Extra floors on small or medium sized buildings which nudge above the tree-line can be similarly detrimental, unless a careful choice is made of positioning, materials, colour and lighting. Indeed, the quality of the built environment that surrounds our parks should be of the highest standards. Possibly even to rival the work of Nash himself.

We recently assessed the potential for new buildings around our parks. This involved, for the first time, examining the contours of the urban landscape, and enabled us to propose some new and valued development. However, other developments must be strongly resisted. Should planning permissions be given for intrusive ugly buildings, or be maintained beyond the lifespan of a building if it is affecting our parks? Why not a planning swap system which encourages developers to build more in less sensitive locations in exchange for preserving the built environment surrounding our green spaces?

Surprisingly, even the construction of monuments and memorials have encroached on our green spaces. Inevitably, because of the very quality of our parks and their historic nature, they are the most desirable location for such structures. However, to maintain our green spaces we should perhaps develop a policy of encouraging such structures to be built in alternative public spaces.

I have discussed here the many responsibilities that the carers of London's green spaces have and their wider roles in helping our communities to re-engage with the natural world. They help make London a pleasant and sustainable city. There are many organisation involved; not just the Royal Parks, but the RSPB, the Wildlife Trust, urban re-generation groups, community gardens, volunteers, children's charities, schools, and many more. Because of the direct and practical role we play, we should have a much more central role in how local and national government and decision makers develop new ideas and strategies to manage our green spaces in the changing environment of London.

Planning and Politics

Chapter 12

The Thames Navigation and its Role in the Development of London

Jacqueline McGlade

THE DEVELOPMENT OF A CITY

Throughout London's history, its centre piece has been the River Thames. At the spot where a road crossed the stream, the Roman invaders of the Claudian conquest of AD 43, displaced the natives and built a bridge and a city — Londinium. At that time, the river was broader and shallower. Just upstream of the bridge (Fig. 1) there were two small but important streams, the Walbrook and Fleet, that flowed into the Thames. They still do so, but through underground conduits. These streams enabled the Romans to bring their boats inside their fortified city and establish foreign commerce. This was the beginning of the city's rich and powerful history (e.g. Ackroyd 2001, Chapter 4 ibid).

Three kilometres further upstream there was another crossing at Westminster. Here there was a small Roman settlement, a religious community in Saxon times, a royal palace in the 11th century, and eventually the growing City of Westminster (see Fig. 2). The bridges spanning the Thames from London to Southwark were wooden until, between 1176 and 1209, a single stone bridge London Bridge was finally built. The tidal waters roared through the 19 arches, and until 1831 "shooting the bridge" in a small boat was one of the thrills of London. A chapel was built on the bridge to St. Thomas à Becket, martyred in 1170, and soon after shops lined both sides. Eventually houses were built above them and water mills installed. By 1300 London had about 80 000 inhabitants that were

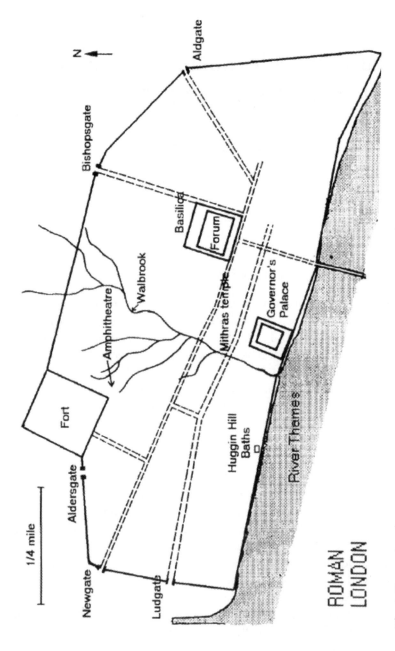

Fig. 1 Roman London.

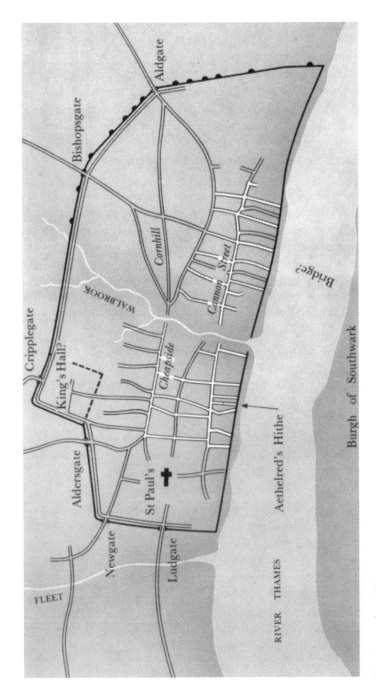

Fig. 2 London 900.

provisioned by a food-supply network extending 65–100 km into the surrounding countryside. The city also drew "sea coal" from Newcastle upon Tyne, 480 km from London by sea.

Embankment of the river seaward of London Bridge began in the 12th century and was completed in the 14th, reclaiming 110 square km of marshland at Rotherhithe and Deptford on the south bank and the Isle of Dogs on the north. But the dynamism of this period came to a sudden end with the outbreak of the Black Death in 1348–1349, with 10000 Londoners being buried beyond the city walls at West Smithfield. Recovery of urban life was to prove a slow process (See Fig. 3).

Towards the middle of the 16th century there was a sudden growth in trade and a trebling of the population in London. Between the commercial city and the royal city, the town houses of nobles and bishops were built, each with its own water gate and eventually a road — the Strand — running from Charing Cross to Temple Bar, one of London's ten gates. But in 1664–1665, the Plague, a frequent invader since the Black Death of 1348, killed 75000 Londoners, followed in the next year by the Great Fire which burned for three days and consumed eighty percent of the City. It was at this time that much of London's contemporary architecture along the river, such as the St. Paul's Cathedral, the Royal and Greenwich Hospitals, and the Royal Observatory of Greenwich, was established by Sir Christopher Wren; indeed there is a Wren inscription in St. Paul's: *Lector, si monumentum requiris, circumspice* (Reader, if you seek a memorial, look about you). See Figs. 4 and 5.

The number and form of river crossings houses also changed over this time. The houses on London Bridge were removed in 1760 when the City gates were taken down, Blackfriars Bridge was completed in 1769, and Southwark Bridge in 1819. Well into the 18th century it was the river rather than roads that served as the main highway, and as foreign trade increased, the land around Rotherhithe and Deptford, which was still marshy centuries later, was excavated for dock building.

Public and private works increased, with the opening of the Metropolitan steam railway line, the building of new Thames bridges and the rebuilding of Battersea, Westminster, Blackfriars and London Bridges. After years of discussion, road bridges outside of the City, passed into public ownership and tollgates disappeared. All the main railways carried their lines northward across the Thames into London, to Victoria, Charing Cross, Blackfriars and Cannon Street stations. The template of the city as we know it today was thus laid down, with the river at its heart.

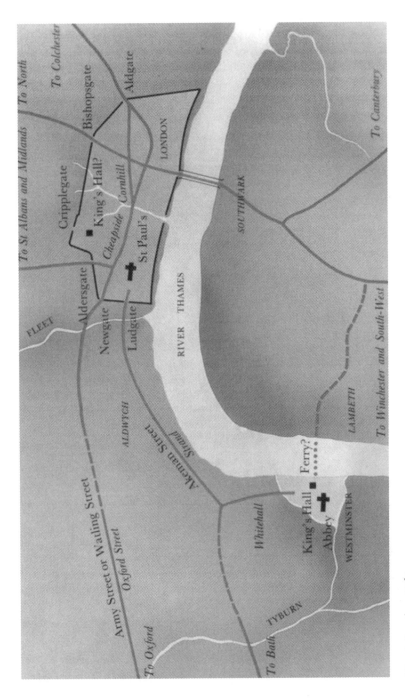

Fig. 3 London 1060.

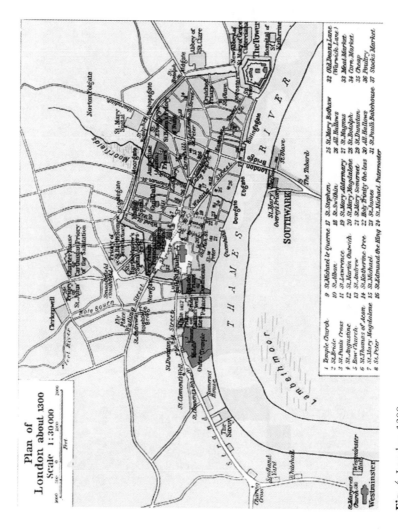

Fig. 4 London 1300.

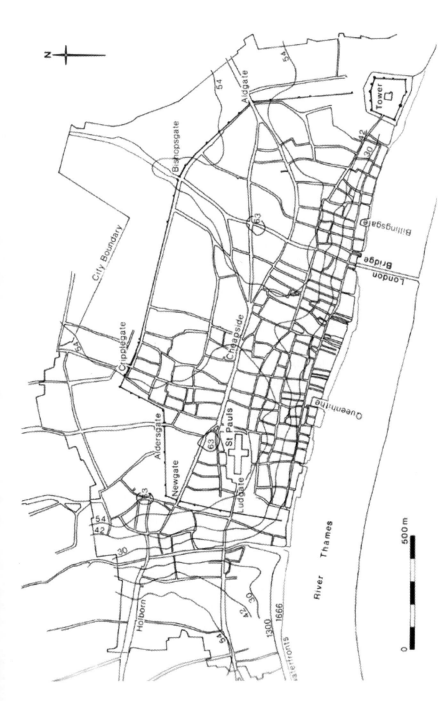

Fig. 5 London 1600.

POLLUTION AND WATER QUALITY

As urbanisation along the river increased, so did pollution. In 1388, Parliament forbade disposition of filth and waste in ditches, rivers and waters, but the act was not very effective. In mediaeval London, latrines were built over any source of running water. The River Fleet, rising on Hampstead Heath joined the Thames at Blackfriars, provided an abundant supply of water, power to drive a series of water mills, and an important navigation route. Upstream the river was seen as a place for pleasure-seeking Londoners attracted by the medicinal waters and tea drinking. But within the city the Fleet was an industrial corridor, characterised by crime, poverty and squalor. In Ben Johnson's 1616 *On the Famous Voyage* he described the foulness of going up the Fleet in graphic terms "when each privies seate is fill'd with buttock? And the walls doe sweate urine and plaisters? ... When their oares did once stirre, belch'd forth an ayre as hot at the muster of all your nighttubs". By 1733 the Fleet became over-whelmed by silt and rubbish, the wharves became roads and the canal was finally covered over.

In the 18th century, wastes from towns throughout the Thames catchment were flowing into the river to be dispersed by the large flow. But problems were beginning to build up from the waste of the 150 slaughterhouses, as well as human excrement. Also toxic waste was discharged from the factories set up in 1807 producing coalgas distributed to gas lights along the street? As Tobias Smollett wrote to Humphrey Clinker in 1771 "... the river Thames, impregnated by all the filth of London and Westminster. Human excrement is the least offensive part of the concrete which is composed of all the drugs, minerals, and poisons, used in mechanics and manufacture, enriched with the putrefying carcasses of beasts and men; and mixed with the scourings of all wash-tubs, kennels and common sewers" (Petts *et al.* 2002). In 1846 the Fleet quite literally blew up — its rancid foetid gases bursting into the street above and sweeping away three poor-houses with the tidal wave of sewage (Trench and Hillman 1984).

By 1855, the population had exceeded 2 million, and in response to the cholera epidemics between 1831 and 1866, which killed 37 000 people, the Metropolitan Commission of Sewers was formed in 1847 to clear the sewers and improve house drainage. Water closets were used by the middle-classes, thereby increasingly overloading the sewer system discharging into the Thames, with regular flushing. The problem was made worse by the mode of flushing out the sewers at low tide, whereby the sewage was held back in the sewers for 18 hours and allowed to stagnate. Sewage discharged at low tide was carried upstream and brought back down on the ebb tide so that during the summer vast quantities of foul mud were exposed along the banks.

Although the creation of sewers improved sanitation, it caused great ecological damage. Prior to the industrial revolution the inner estuary below London Bridge was a wilderness of marshes and reedbeds, with bitterns, bearded reedlings, spoonbills and montagu's harrier common inhabitants. And the river had important fisheries of salmon, smelt, shad, flounder, eel and whitebait, plus cultured mussels and oysters. But by 1859 William Yarrell wrote that smelt had become uncommon and "the last salmon I have note of was taken in June 1833".

The drive to finance major new building schemes was slow, partly because of a lack of political will, but also because of a dearth of scientific understanding about the link between waste water and public health. It took the Great Stink in 1858, when the Thames stank so badly on July 3 that Parliament was nearly closed by the "pestilential odour". Sir Joseph Bazalgette was given the task of designing and building three lines of sewers to intercept the existing sewers and deliver the sewage to outfalls 17 km below London Bridge. Bazalgette's system was based on gravity, using pumping stations at various points to raise waters from the lower-level sewers to meet the higher-level northern and southern interceptor sewers, leading to the treatment works at Becton and Crossness. By 1865 there were nearly 2000 km of sewers and a main drainage works, which had cost over £4 million.

Damage to the Bazalgette system during the two World Wars, together with overloading of the sewage system despite new treatment and plant, meant that by the 1950s the Thames had deteriorated to the extent that the water contained no dissolved oxygen in the summer, killing virtually all wildlife in the river. A survey of remnant marshes between 1949 and 1950 found few wildfowl upstream of the sewer outfalls, and a survey of fish in 1957 showed that there were no established fish populations between Kew Bridge and at least as far downstream as Gravesend, 50 km away. Indeed, studies in the 1960s showed that the Bazalgette system actually made the situation worse by concentrating inadequately treated sewage which was becoming trapped in a reach 5 km below the discharge point.

In the second half of the 20th century, a better understanding of the tidal flows of the river and the dynamics of the Thames Basin enabled new processes to be added, such as advanced filtration and aeration to improve the quality of waste water discharges. (Chapter 10) Solid matter was also treated to render it non-toxic. By 1975, the small inefficient works were all but gone and summer oxygen levels, which in 1963 were close to being anaerobic for 18 km below London Bridge, had recovered, and within 10 years fish and birds had largely returned to the Thames. The species most obvious to benefit were the migratory fish and birds such as smelt and salmon, and the winter populations of pochard, shelduck, pintail and tufted duck. The upper tidal river became once again a refuge for water birds, with 10 000 wildfowl and 12 000 waders.

MANAGEMENT OF THE THAMES AND LONDON'S RIVERS

The Thames runs for 345 km from its source in the Cotswolds to its marine extent at Southend: 9873 km² of southern England are drained by the Thames and two thirds of London's drinking water is supplied by it. From its source at Lechlade, the Thames flows through gently rolling pastoral lowlands, with an average fall of less than 32 cm per km. At the other end of the river, the picture is quite different with tides and surges of the sea having a profound effect on the water level of the river's lower course. Straightening of the shoreline and jacketing the river with stone walls moved the tidal limit from London Bridge 31 km upstream to Teddington Weir. This tidal influence can be seen for some three hours during a high tide, at Teddington, although the transition from freshwater to estuarine waters occurs closer to central London, around Battersea. At London Bridge, in the heart of the city, the river rises 7 m on the spring tides and 5.5 m on the neap tides. It is now known that the average high tide is rising at a rate of about 1 m each century, as London sinks: high tide at London Bridge is 1.2 m higher than it is downstream at Southend.

The average flow at the upper limit of the tideway at Teddington, is 53 m³s⁻¹, rising to 130 m³s⁻¹ after the winter rains. In extreme floods (e.g. in March 1947) the discharge at Teddington Weir can be as much as 590 m³s⁻¹. Reputedly, an average of 887 m³s⁻¹ passed over it in one day after great storms in 1894. The river in spate can upset tidal flows for some miles below Teddington, overpowering the incoming tide and causing the stream to run seaward continuously for days on end. Conversely, high spring tides can overtop the weir and affect the river flow as far as 3.2 km upstream of Teddington.

Traditionally, land drainage and flood management practices involved the straightening of watercourses within concrete channels, culverting and excessive flood protection measures. These have left a sterile legacy of rivers across London. Within the Ravensbourne catchment in south London, for example, the Environment Agency has identified that 37% of the rivers are culverted and a further 25% constrained within vertical walled channels. The policy in river managers to build culverts and straighten channels arose because they had to cope with increasing flows, arising from new and enlarged impermeable areas, of building development. Today, this is no longer seen as environmentally acceptable and the Environment Agency now operates a policy against culverting. In fact rivers are being restored using more natural bed forms with pools and riffles. Then ecological networks are created. Natural banks are extended with gabion baskets and willow spilling instead of concrete. Also channels are split into two smaller watercourses, each having a lower flow rate. Excess flows may spill over into surrounding grassland, which helps improve the biodiversity of the whole area, and creates a greener landscape.

A good example of such enhancements can be found in the regeneration of Wandsworth Riverside; here the tidal flood defences at the mouth of the Wandle are being improved to include inter-tidal terraces along 60 m of the river wall at its confluence with the River Thames. This is creating high quality habitats for invertebrates, fish, birds and bats not currently found in the vicinity. The site also provides one of the most important locations along the Thames for wading birds. To minimise the threat of disturbance a number of barges are being moored within the river to offer birds an alternative roosting habitat.

Further down river on the Greenwich Peninsula this type of thinking led to the creation of a new design for flood defences along the 1.24 km river frontage of the Millennium Dome. Here there an extensive inter-tidal habitat, boardwalks for public access, a series of salt marsh terraces, and the conversion of many redundant structures into 'green piers'.

Nevertheless the hazards of flooding along the Thames remain. Recognised as early as the 12th century when the construction of tidal defences began, the law of the marsh, established in Essex in 1280, required every man to contribute to the upkeep of defences in relation to his land rights and benefits. In 1427, a parliamentary authority was established and reinforced by an Act in the 16th century which established the principles of land drainage and tidal defences for the next 300 years. A series of increasingly damaging high tides between 1874 and 1877 led to the passing of the Metropolis Management Amendment Act in 1879 (Thames River Prevention of Floods). This led to the building and subsequent raising of defences as surge tides continued to exceed defence levels. A surge tide in 1928 caused some people to be drowned in London. This led to the 1930 Land Drainage Act which was to remove drained waters, place financial responsibility with river catchment boards and set statutory defence levels at 5.54 m at Hammersmith, 5.28 m at London Bridge and 5.18 m at Barking Creek.

During the 1953 surge tide when Canvey Island on the Thames Estuary was completely submerged, more than 300 people died, 25 000 properties were damaged and 30 000 people were evacuated. This led to further research into the catastrophic potential of tidal surges for London's underground infrastructure, buildings, and population, and prompted the construction of the Thames Barrier at Silvertown. This was completed in 1982 and spanned the Woolwich reach (Gilbert and Horner 2002) with extensive complementary flood defenses along the entire tideway.

Today, the Environment Agency has responsibility for the flood defences along the Thames and its tributaries. These include 400 structures used for controlling the flow such as locks and weirs, 337 km of river walls and embankments; there are five barriers at Barking, Benfleet, Dartford, Easthaven and Fobbing Horse, and 30 industrial flood gates at the entrance of the Tilbury and Royal Docks. Overall, London and the Thames Estuary are designed to be protected from flooding even for the worst storm that is likely in the next 1000 years, based on past statistics. There is some additional allowance for a further rise in tides and in sea level. The Thames Barrier has been in operation for 20 years, with on average 3.3 closures per year. However the number of closures is expected to rise in relation to changes in storm surges, sea-level rise, and sinking of the land. In 2001 there were 24 Thames Barrier closures and in January 2003 the exceptional rainfall led to 14 closures on consecutive tides to alleviate flooding at the tidal-fluvial limit of the Thames. It is likely that in these changing conditions, the flood defences will have to be strengthened (see Chapter 13).

THE THAMES LEISURE ECONOMY

The Thames has always played an important role in London's leisure economy. During the 17th and 18th centuries, frost fairs on the frozen Thames were a common occurrence. The development of the sewerage system was accompanied by the creation of large open spaces along the river known as embankments, which became the focus for the growing interest in pleasure gardens and river cruises. From 1864 to 1870 the main sewers were laid along the north bank behind stone retaining walls and roadways laid over them from Blackfriars Bridge to Westminster. About 15 hectares of mudbank were reclaimed and converted into parkland — Victoria Embankment Gardens between Waterloo Bridge and the Hungerford Railway Bridge. In 1874 Chelsea Embankment, also laid over new sewer mains, was completed, so that the embankment system, interrupted briefly by the Houses of Parliament, became seven kilometres long. On the riverside opposite, the Albert Embankment was completed in 1869 as a flood-control installation, and extended for pedestrians only, in front of what was London's County Hall, in 1910. The last of the embankments was made in 1951, when the South bank cultural complex was started on a bombed out site between Westminster and Waterloo Bridge.

In the 235 km above Teddington, where the Thames is non-tidal and fully freshwater, there are 44 locks, 109 leisure and recreation clubs and space for up to 26 000 boats. These activities generate a total annual income of £85 million. Also the river is host to 14 million leisure day visits, and 28 million casual local visits each year, generating £119 million of expenditure each year and supporting 72 000 jobs in the tourist industries in the riparian districts. Along its freshwater reaches where there are 12 000 houses sitting within 500 m of the river, proximity to the Thames adds about £580 million to the value of these riverside properties. Along the river stretches above Teddington there are 65 Sites of Special Scientific Interest (SSSI) and 60 km length of riparian Areas of Outstanding Natural Beauty. There are flood meadows for rare plants such as snakeshead fritillaries and green-winged orchids, Lodden lilies or summer snowflakes and 30 different species of fish.

Below Teddington the tidal river is one of the largest open public spaces for recreation and leisure in the capital. As well as environmental designations, there are three world heritage sites along the river — the Palace of Westminster, the Tower of London and Maritime Greenwich. In 2002, twenty-five cruise liners visited London generating £6.3 million of visitor spending; and each year 3 million people travel on the Thames by boat.

London's river corridors form the backbone of the open space framework across the capital, with 800 000 people living within 10 minutes walk of a canal or river. In planning and designing new developments it is necessary to understand which areas must be enhanced for leisure activities or protected for their biodiversity and wildlife. In either case decisions also have to be taken about which species to introduce or reduce. In recent years a number of forms of enhancement have begun to emerge, including the use of buffer zones which encourage the incorporation of green strips alongside watercourses. A good example can be seen in Brentford where the Grand

Union Canal, which in the 18th and 19th centuries provided the main freight transport link between the Midlands and the River Thames, is being regenerated. Local housing schemes have incorporated new wetland areas, new tidal habitats for species such as the Two Lipped Door Snail (a rare species in the red data book). The regeneration has improved the overall visual quality of the canal area, as well as expanding public access and mooring facilities.

And further down the tidal region, there are estuary sites of critical environmental significance that have been designated under the EC Habitats and Birds Directive, including the North Kent Marshes Benfleet and Southend Marshes (Special Protection Area, Ramsar site and Site of Special Scientific Interest (SSSI)), the Thames Estuary and marshes including Mucking Flats, the Medway Estuary and marshes, all of which are being carefully protected and conserved.

THE COMMERCIAL RIVER ECONOMY

The development of the docks below London Bridge began during the reign of Queen Elizabeth 1, when legal quays were established on the north bank between London Bridge and the Tower of London. These quays were the only place where ships could land dutiable goods legally. The traffic soon grew to be so large for the quays that in 1663 Parliament allowed the establishment of alternative "sufferance wharves" on both banks. Port activity doubled between 1700 and 1770, and by the end of that time, the Upper Pool — the part of the river that stretches for a kilometre below London Bridge — held as many as 1 775 ships in a space allocated to 600. Cargoes unloaded into lighters to be taken to the wharves were sometimes caught on the river for weeks at a time and became vulnerable to pilfering and raids by river pirates.

To enable vessels to offload into guarded quays the West India Docks were opened in 1802 at the north end of the Isle of Dogs, a marshy river peninsula opposite Greenwich Hospital. In 1805, the London Dock opened in Wapping and downstream from the Isle of Dogs the East India Dock was inaugurated in 1806. In 1807, the existing Greenland Dock (where blubber was rendered) was combined with the Howland Wet Dock in Rotherthithe on the south bank, and later enlarged to the Surrey Commercial Docks, covering 185 hectares. St. Katharine's Docks were built under the lee of the Tower in 1828.

In 1909, the Port of London Authority (PLA) was created to take over ownership of all existing docks and control of the river and the port. But as volume of river traffic fell, the PLA set about closing down the upstream docks, and in the 1960s decided to sell its riverfront properties covering 300 hectares. In 1969, the Greater London Council bought St. Katharine's Docks, and within three years its value had quadrupled. Some of the warehouses were retained, and a theatre, marina, restaurants and pubs were built as part of a village for 2000 residents occupying houses, flats and studios. This was followed by regeneration of waterfronts at Thamesmead on the Royal Arsenal of Woolwich, the Royal Victorian Victualling Yards, and Deptford. The length of waterfront from Waterloo to Woolwich, formerly declining docks, factories and decayed housing, became the focus of the Docklands development — a time of frantic bidding and fevered planning by the newly created London Docklands Development Corporation. The outcome has seen not only iconic buildings such as Canary Tower and the Millennium Dome, but the cleanest metropolitan river in the world.

The river continues to be an important conduit for commercial traffic as well as for commercial development along its banks. For example, of the 50 million tonnes of cargo handled by the PLA during 2000, 10.5 million tonnes was handled within the Greater London limits with 750 000 tonnes of London's household waste being transported along the Thames.

A MIXED FUTURE

The draft London Plan, prepared by the Greater London Authority (Mayor of London 2002) sets out a whole range of policies of relevance to the leisure aspects of the river: these include the Blue Ribbon Network, floodplains, flood defences and sustainable development. London's Blue Ribbon Network refers to the Thames, the canal network, the other rivers and streams within the city and open water spaces such as docks, reservoirs and lakes, and is of strategic importance as a resource for transport, leisure and tourism and as a principal component of London's public realm.

The proposed policies stress the importance of the water economy to London and include strengthening the linkage between urban regeneration and waterways. Waterways are to be improved, mixed use housing and industrial developments are to be promoted, especially when the industries use water transport and have minimal adverse environmental impact. The use of water-space by land-based activities needs to be planned so as to benefit the waterways. For water-spaces to provide an aesthetic experience, water-views need to be protected, along with local character, heritage sites and archaeology. These spaces are ideal for exhibiting art in many different forms. The policies recommend new initiatives for passenger and freight transport, for tourism, and leisure and marine support facilities. This is only possible by safeguarding and protecting of wharves. Lorry movements associated with building operations constitute a significant volume of road traffic, this could be reduced by more extensive use of river transport for this purpose. Carefully designed new river crossings need to be planned, and perhaps the removal of those that are defunct. Conservation policies include resisting loss of habitat and species diversity and creation of new habitats by resisting culverting and impounding of rivers and streams. For water risk and resources management there should be identification of risks of flooding in any riverside developments, encouragement of minimisation of water consumption, and protection and improvements to water quality. Residential moorings should be an integral part of the environmental planning since the local inhabitants take an intense interest in every aspect of the river and often act as catalysts for improving the environmental health of the riparian zone.

These policies should ensure that any new development within and alongside the water space would, generally, only be allowed where it was strictly necessary, and where habitat value could be increased and flood risk adequately taken into account. So for example, housing, roadways, etc, proposed for the Thames Gateway would need to take into account not only the current tidal fluctuations in the estuary, but also the continued rise in sea level caused by global warming and geology. The designs for development in such areas at risk would therefore need to consider not only the requirement for flood defences and extended warning systems, but also the impact of any local water shortages that might result from increases in local population and rising peak temperatures in the summer. Improved drainage regimes would also help manage long-term flooding risk.

Land alongside the river's open spaces should be concentrated on activities which specifically require a waterside location, such as water transport, leisure and recreation. The sustainable distribution of goods and services is leading to a renewed interest in using the waterway network to move goods and people around, thereby reducing congestion and minimising the environmental effects of heavy goods movements. Then new cruise facilities, piers and dedicated stopping facilities on the canals and river will be needed. Plans for these areas should also prevent the removal or redevelopment of support facilities such as boat building and repair, mooring sites, boat houses and slipways. Wharves with existing or future potential for handling aggregates and other cargoes need to be safeguarded, including their road access. Despite river freight taking thousands of lorry journeys off of London's arterial roads, Transport for London is unable at present to fund improved road access to wharves. Also no special consideration for road access is given to those who live near wharves.

It is stated in the Plan that waterside developments also need to respect the particular nature of the river and its tributaries and adjacent canals by preserving views and promoting safe public access. In other words the Thames should not just be developed as a private resource or backdrop which only privileged people can afford to enjoy.

Without doubt the Thames is a highly complex mosaic of environments and uses. In considering its management, it is useful to divide the river into two parts: the non-tidal above Teddington and the tidal below. Indeed there is a clear change at Teddington, separating the upstream "Wind in the Willows" character of recreation and boating from the busy, commercial and tourist waterway that flows through central London and down to Tilbury and beyond. Recreational use, with rowing and sailing boats, continues in this stretch for 15 kilometres downstream of Teddington as far as Wandsworth. Below Twickenham (which is roughly 8 kilometres downstream of Teddington) the river changes dramatically, with flow rates of 5 knots in and out for at least half the day. From Wandsworth down to Westminster the level of recreation is very low, and drops off completely through Central London, where tourist vessels tend to dominate the waterway. Below Tower Bridge recreational and sporting use picks up, with famous rowing clubs based there and a lot of sailing craft.

Those managing the Thames have their own vessels. Different types are used in the upper and lower reaches of the river, where different safety protection methods have evolved. Above Teddington, the river only flows in the downstream direction and the currents are much lower than further down. Here safety cover takes the form of patrols from the bank side and the presence of lock-keepers combined with boats provided by the Environment Agency. The PLA safety patrol vessels get up as far as Teddington, but only on part of the tide, because at Chiswick and Old Isleworth there is hardly any water at Low Water. This is why draft vessels find it difficult to navigate upriver from here. From Tower Bridge down to Greenwich there is a mixture of tourist and recreational vessels together with freight movement.

Thus, if there is to be a system of safe and quick commuting transport along and across the river, a number of bodies must co-operate effectively. Moreover, for passenger transport to be viable on the Thames there needs to be sufficient numbers of piers enabling people to cross the river as well as move along it. At present, passenger transport stretches from Tate Britain in the west to Great Eastern (Isle of Dogs) in the east. New piers have been constructed at the Millennium Dome and at Woolwich, but these are too far apart to play much of a role in day-to-day transport, and Greenwich pier is operating at full capacity. Clearly, there is also a need for new piers on both sides of the river between Deptford and Woolwich.

People have been brought back to enjoy the Thames. This has been one of the key benefit of recent riverside developments ranging from the Millennium Dome and the Greenwich Peninsula to the new "wobbly" bridge crossing at Tate Britain. To a large extent the development of London over the past three decades as a world city has benefited from its 'blue-water' space. The question now is whether or not the future dynamics of the river, with the increasing risks of flooding from sea level rise, elevated tide levels and surges, and altered patterns of rainfall, means that the need to protect the billion-pound investments of London's infrastructure will necessitate such high walls and riverside flood defences that the Thames will end up by being closed off from the public eye. Whatever the solution, there is no doubt that the River Thames will continue to play a major role in the development of this world city.

REFERENCES

Ackroyd, P. (2001) *London, the Biography*, Viking, London.

Gilbert, S. and Horner, R. (2002) *The Thames Barrier*, Thomas Telford Ltd., London.

Mayor of London (2002) The draft London Plan. *Draft Spatial Development Strategy for Greater London,* Greater London Authority.

Petts, G., Heathcote, J. and Martin, D. (2002) *Urban Rivers our Inheritance and Future*, IWA.

Publishing and Environment Agency, ISBN 1-900222-22-1.

Trench, R. and Hillman, E. (1984) *London under London: A Subterranean Guide,* John Murray Ltd., London.

Chapter 13

Dealing with Disasters

Dennis J Parker and Edmund C Penning-Rowsell

INTRODUCTION

London cannot be a sustainable city unless its hazards and disasters are effectively dealt with and the security and health of today's and tomorrow's Londoners are protected. This requires managing hazards and disasters so as to limit people's exposure to hazards while balancing this objective against other social and economic goals and benefits. It is essential to take a holistic perspective which recognises that any form of change in one part of the complex 'London system' may well generate increased natural and other kinds of risks elsewhere in the system. Managing hazards effectively also means investing in comprehensive scientific understandings of the causes of hazards and disasters, and relating this to lessons learned from disasters here and elsewhere around the world. It means pursuing strategies which increase hazard and disaster resistance and resilience, and integrating hazard management into all aspects of London's governance and development (Parker 2000). Without the participation of Londoners, hazard management can only be partially successful (see Chapter 15). Therefore we must seek to foster understanding of management strategies by the public and their participation in the resulting decision. These are some of the key issues that lead to questions about the extent to which we are adopting effective hazard and disaster management strategies. These need to be sustainable in the context of a world city like London.

This chapter begins with an examination of some of the main features of mega-cities, of which London is a prime example. As mega-cities develop their hazards and their management are being transformed as we see in the way that London is dealing with the various types of flood hazards. A risk assessment for Londoners is proposed, which leads to the conclusion that there are real deficiencies in the management strategies to date. Hazard management in London could be improved on a realistic time scale at a realistic cost.

THE MEGA-CITY PERSPECTIVE

We are living at a moment in history when the world has become predominantly urban. Furthermore the world is becoming dominated by very large cities — 'mega-cities' with a population of about 8 million or more. Mega-cities are a comparatively recent phenomenon, especially in the developing world where they have emerged with the greatest rapidity (Soja 2000). Knowledge about how mega-cities interact with their natural environment to create hazards and disasters is comparatively limited. There is much more knowledge based on the experience of small to medium sized cities and it is not yet clear whether mega-cities are experiencing more, less or the same levels of hazard and disaster vulnerability (Mitchell 1999). However it is clear that mega-cities vary in their vulnerability, and also that there are some ominous signs for the future for all their inhabitants including Londoners. Since 1939 over one hundred natural disasters have affected large cities, e.g. the earthquakes in Tangshan in 1976, Mexico City in 1985, Cairo in 1995, and Kobe in 1995; the hurricanes and windstorms which affected Miami/Dade County in 1992, and London in 1987. The number of mega-cities located in the coastal zone with its risks of subsidence, flooding and cyclones increased from 2 in 1950 to 13 by 1990, and is predicted to be 20 by 2010 (Nicholls 1995). Sea-level rise induced by global warming is progressively increasing this exposure. It is becoming clearer that very large cities have enormous potential for disaster creation and many appear to becoming more vulnerable. This is due in part to the pace of development which often defies control in hazardous zones, and to the large impacts that huge concentrations of people, industry and transportation have upon biological and physical systems which are being stretched to their limits. As the infamous September 11th disaster in New York demonstrated, unfortunately, large concentrations of people and investment also appear to be increasingly attractive to terrorist and cultist groups who wish to create large losses intentionally. Londoners are no strangers to this particular trend.

As the nature of hazards and disasters transform in time and space, they require new thinking about how they can be anticipated and successfully managed. In mega-cities, and perhaps elsewhere, distinctions between 'natural' disasters and ones which may be described as 'technological' or 'industrial' are becoming increasingly blurred. At the same time, it is increasingly difficult to separate these types of disaster from biological and social ones. This is particularly so in urban settings where the sources of risk are multiple and where the experience is often of an interactive mix of hazards (Fig. 1). The flood which damages a rail line, de-railing a train carrying fuel and inflammable material causing an explosion and pollution of land, air and watercourses is one example. The flood disaster that has as one of its effects the contamination of drinking water supplies is another. In London this and other hybrid hazards such as the combined risks of flooding and crowd disaster in the underground have to be planned for. Severe weather and windstorms may shut down transportation systems owing to loss of power, in turn leading to crowd handling crises and other 'spin-off' emergencies. The range of combinations and possible

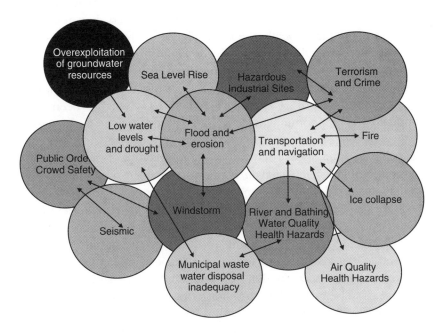

Fig. 1 Hybrid hazards in London.

disaster scenarios is particularly large in very large cities and has been exercising London's emergency planners for some time.

THE LONDON MEGA-CITY AND ITS HISTORICAL EXPERIENCE OF HAZARDS AND DISASTERS

London is one of the oldest of the contemporary mega-cities. In about 1790 London became the world's first post-industrial city to reach a population of one million, and in so-doing became the first embryonic mega-city. Since then, and depending on the various ways in which London's boundaries may be defined, its population has risen to between around 7 million to 13 million or more. London is not as intensely hazard-prone as some of the world's megacities, such as Tokyo or Los Angeles, but its history is nevertheless punctuated by a long series of disasters. Today Londoners live and work in a context of ever-present, multiple hazards.

Longevity Leads to Learning Opportunities

London's longevity as a very large city affords a special perspective upon urban hazard and disaster management. Events of very low recurrence interval (e.g. the Black Death, the Great Plague of 1665, the Great Fire of 1666, the Second World War blitz, the sinking of Princess Alice in 1878, the Thames tidal flood of 1953, the windstorm of 1987, the smog of 1952, the Bishopsgate bomb of 1993 and the Kings Cross fire of 1987) show up in the historical record of experience. There has therefore been correspondingly more opportunity over time for social learning of strategies for making Londoners resilient to a wide range of environmental extremes. Age also imparts a special flavour to the appraisal and management of environmental risks, reflected in London's case for concern about safeguarding historical buildings and places, and irreplaceable art objects which are part of Londoners' cultural heritage. As in other ancient cities, the longevity of London leads to some indifference by public and metropolitan authorities as to the vicissitudes of nature. The attitude is certainly displayed in editorial pages of serious newspapers. Taking the long view, it may be argued that London has learned from experience and has proved itself to be one of the most disaster-resistant of mega-cities. On the other hand there is evidence that lessons have not been learned as well as might be expected. On balance, and to be a little controversial, we tend towards the latter conclusion — a conclusion argued in greater detail than is possible in this paper in Parker (1999).

London and Londoners as Witnesses to Transforming Hazards

London has witnessed a transformation in the nature and perception of hazards over time. Environmental hazards played an important role in London's evolution, and throughout much of the city's existence the exposure and vulnerability of individuals has been differentiated by social class and wealth. Affluent classes sought to reduce their exposure by moving out of places that were naturally or socially at risk. In the process they gradually created exclusive neighbourhoods (for example, to the west), characterised by services and amenities that did not exist in lower-class districts. Compared with later periods, little is known about the hazards of medieval London but they are widely believed to have been numerous. Clout and Wood (1986) portray a society whose residents were intimately acquainted with threats to survival. Densely-packed timber buildings frequently burned down, drinking water was badly polluted, epidemics abounded, and mortality rates were high. Destruction wrought by fire provided a major opportunity to redesign London in the 17th century, with regulations covering the width of streets, the height of buildings and the use of fire-proof construction materials.

By the time of the first official census (1801), London's population had risen to 1 117 000. The city's accelerating growth was tied to Britain's emergence as a colonial trading nation. Shipping of goods through London greatly increased creating demands for new docks which formed the basis for society in the East End. This was characterised by insecure employment, poverty, inadequate housing, immigrants without adequate material resources and disease. Although, at the height of the British Empire, Victorian London was 'the most commanding concentration of people, trade, industry and administration to be found anywhere on the globe' (Clout and Wood 1986), it was also a scene of many hardships: deprivation, poverty, pollution, squalor and disease. At the end of the 1850s a series of ultimately successful hazard reduction measures gathered momentum (e.g. abstraction of water from the tidal reaches of the Thames was banned). Air pollution was a persistent hazard in London from the 13th century onwards. Pollution episodes were closely linked to trends in urban and industrial development as well as to variations in climate. Attempts to alleviate the problem were never wholly successful and air pollution remained a major contributory cause of death and disease amongst Londoners, especially in the 19th century, but also as late as the 1950s. During the 1939–1945 war London became one of the first large cities to experience sustained aerial bombardment, with the impact being comparable to that of a great earthquake. Most systems continued to function; electricity, water, rail etc., because they were quite decentralised and resilient.

How Policies to Protect Open Space are Linked to the Generation of Hazards

Since the end of the Second World War, the hazard potential of London has been strongly — if indirectly — affected by public policies that were aimed at limiting the city's size, controlling the distribution of industry and revitalising the decayed inner areas. The susceptibility of London's concentrated population to aerial bombardment, the declining quality or urban life in a sprawling city choked with traffic, and a rapidly disappearing countryside led London to adopt a statutorily protected open-space around London called the green belt. The temporary war-time evacuations helped break established patterns of urban life and opened the way for post-war voluntary migrations to the outer suburbs, and the Abercrombie plan led to relocation of many Londoners to new and enlarged towns beyond the green belt, allowing blitzed and blighted London to be redeveloped. Although the physical growth of London was halted, its employment growth continued so that by the 1960s the highest rates of housing and population growth were 55–110 km from the city centre and mass commuting had become established.

The effect of the green belt on the quality of life and exposure to hazards has been complex. The city's centripetal growth, itself hastened by green belt policy, has partly alleviated the build-up of further environmental hazards in the metropolis by reducing concentrations of people and investments at risk. Conversely, and at the same time, there has been an intensification of environmental problems beyond the green belt, including encroachment of building and infra-structure development onto floodplains (except those which are also green belt) (Burby *et al.* 2001). As car ownership has mushroomed, motorways developed and rail services have declined in attractiveness, a much larger area has become afflicted with traffic congestion. Therefore many of the London region's environmental problems have been exacerbated by the process of decentralisation. Traffic congestion, waste disposal, poor air quality, noise and flood hazards have become significant environmental and political issues across the southeast region.

The Unconsidered Relationships between Development and Increased Hazard Potential

Containment of London's growth contributed to rapid development along the M4 corridor and the 'western crescent' just beyond the green belt during the 1970s and 1980s in particular. Here 'high-tech' electronics companies became established. Development in these parts of the Thames catchment increased the rates of water run-off and contributed towards worsening flood hazards. This was further exacerbated as the Thames and Thames tributary floodplains witnessed a build up of residential development. Here the influence of urban development on the growth in hazard potential were rarely taken into full consideration. A case in point was the regeneration of London's Docklands during the 1980s. The redevelopment of Docklands was a reflection of a private property boom that occurred throughout much of London during the second half of the 1980s, aided by the deregulation of British financial services in 1986. However, strategic planning for this redevelopment did not adequately consider the impacts of new construction upon traffic congestion and pollution, nor the potential for increased exposure to other environmental hazards such as flooding.

How Londoners Experience Hazards

Londoners experience many minor hazards in their daily life. These are punctuated by close proximity to (and sometimes involvement in) accidents, hazardous situations and occasional disasters. The experience is both personally witnessed and felt, and also received through intense media coverage and speculation. How far Londoners are actually vulnerable to hazards is in part related to their standard of living which varies across the city. Since the 1970s London has experienced a marked increase in income inequality and social polarisation that partly differentiates Londoner's experience of, and vulnerability, to hazards. Not only have the better-off grown more wealthy but the worse-off have grown relatively poorer. Racial and ethnic minorities have been disproportionately affected by these trends because they are heavily concentrated in London, particularly in the inner city. The inhabitants of boroughs such as Islington, Hackney and Tower Hamlets suffer from a range of urban ills including overcrowding, problems associated with rental housing, homelessness, lack of access to cars, severe unemployment, noise and crime. Gentrification in adjacent areas sharpens the experience. Many residents, and not only poor ones, express growing disillusionment with the quality of life in London, especially in the inner areas. The problems are exacerbated by old infrastructure, inadequate maintenance, poor ambulance call-out response times and the high costs of living relative to other UK regions (see Chapter 6).

MANAGING THE TIDAL FLOOD RISK

The Progressively Increasing Risk

Tidal flooding presents London's most serious flood risk: without existing protective measures tidal flooding would be common and impacts would be potentially very large (Doyle 2003). The risk of tidal flooding is steadily increasing. Mean regional sea level has been rising at about 0.36 metres per century, but the rate of rise is much faster in some places (e.g. 0.8 metres per century at London Bridge). Twice in 1978 the Thames reached levels similar to those of the highly unusual catastrophic floods of 1928. Several factors are responsible for these trends. They include geological processes, global warming, and human modifications of the hydrological system. For example, southeast England is gradually sinking to compensate for the continuing isostatic rise of northern Britain that began with the departure of the last glaciation. Subsidence has also been occurring in areas of excessive groundwater extraction. In addition, sea levels are rising because of global warming. The cumulative effects of dredging, embanking of former marshlands along the Thames, and encroachment of buildings and embankments into floodplains around the Thames have also increased the tidal range.

There is also a very low risk — but one which is difficult to quantify with accuracy at the moment — of catastrophic and sudden change in sea levels as a consequence of the failure of the Antarctic ice shelf (Intergovernmental Panel on Climate Change 2001). There might be changes in sea level of between 3 to 6 metres. Such changes would lie well outside of the envelope of current tidal flood risk strategy for London and would pose major new problems.

Evolving Tidal Flood Risk Management Strategies

The traditional response to London's tidal flood hazard has been piecemeal and incremental. Flood embankments were raised by small amounts to keep pace with increasing threats. By mid 20th century the limits of this strategy were recognised and new approaches were considered to be necessary.

Because of the increased frequency of high tide surge levels, in the early 1980s the Greater London Council mounted a major public flood warning awareness campaign, and a flood emergency plan for London was developed. The public campaign comprised of mass distribution of flood leaflets to households in vulnerable areas, posters in prominent places, and graphic television advertisements. Unfortunately, on the rare occasions when awareness of flood warnings was subsequently investigated, the results were not encouraging. Surveys revealed that after receiving a one-hour warning many Londoners planned to take inappropriate actions, such as seeking to use the underground system (Penning-Rowsell *et al.* 1983). Since the Easter 1998 floods which affected parts of the upper Thames catchment tributaries and the Midlands, the Environment Agency has learned much about how to successfully raise public awareness of flood-ing, flood warning and appropriate actions to take on receipt of a flood warning, and current knowledge applied to London's tidal flood risk could almost certainly improve on the results of the 1980s campaign (Bye and Horner 1998; Environment Agency 2001).

The flood warning campaign was viewed as a stop-gap measure that would buy time pending the construction of a major flood barrier across the lower Thames, which was finally opened in the early 1980s. The Thames barrier is the centre-piece of a major flood defence system which extends downstream of the barrier for many kilometres on both banks of the Thames. The Barrier was designed to protect against extreme tidal surge events at least until the year 2030 (Gilbert and Horner 1984) — a date which is now rapidly approaching. On receipt of a warning from the storm tide warning system (established after the 1953 floods), gates can be raised off the bed of the river into a closed position. The Barrier was designed not only to contain the estimated 1000 year flood but to take into account of a continuation in the upward trend of high-water levels plus an additional 1.5 m safety margin. The top of the barrier is 7.20 m above the local mean sea level. Downstream embankments have been raised, and there are other smaller flood barriers on tributaries. One potential flaw in the system is fail-ure to take account of the possibility that global warming may accelerate the pace of sea level rise. A faster rate of rise would reduce the effective life of the barrier. It is difficult to assess the added risk because of the uncertainties about global warming. Several sea level rise scenarios were considered by Turner *et al.* (1990). The most likely case assumed a net rise of 0.31 m between 1990 and 2100, whereas the worst case assumed a net rise of 1.1 m during the same period. Under some sea level rise scenarios, the probability of flooding may be 2 to 10 times more likely than at present. Whatever the scenarios considered it is likely that by 2030 the 1000 year flood protection afforded by the Barrier will have declined to about half or less of its effectiveness when it was commissioned.

Dramatically Increasing Flood Exposure Potential

The area protected by the Thames Tidal Flood Barrier now contains approximately 1 million permanent residents and thousands of commercial properties. If daily commuters are included the number of those affected by flood-risk rises to around 1.5 million. Paradoxically, as London's protection against tidal flooding has improved, its potential exposure to tidal flooding has actually increased. The existence and successful operation of the Barrier flood defence system since the early 1980s has lead to the widespread public and political belief that the risk of flooding in the tidal area has effectively been eliminated. To even admit that there is any post-Barrier flood risk in London is a sensitive issue in some circles of government. At times public debate has been deliberately limited. Yet all hazard management requires a balance to be struck between economic and social benefits of development and the natural and other risks involved, and this balance is best struck through dialogue with the stakeholders.

The development and redevelopment of London has had major implications for flood hazard exposure. The facts are that the Thames floodplain is the site of one of the largest increases in flood loss potential in Europe (i.e. the expected flood losses would be very large indeed, as a consequence of rapid redevelopment, without the protection afforded by the Barrier defence system). Without the Barrier defence system in place, tidal flooding would be highly probable with high consequential damages. Indeed financial losses would be catastrophic to the economy of London.

The tidal flood-prone area includes nearly the whole of London's redeveloped Docklands and riverside areas between the City's financial district and Woolwich, both of which were the sites of major investment from the 1980s onwards. A number of new developments have been designed to minimise direct flood damage potential (for example, by elevating the ground flood levels) thereby integrating flood risk management into redevelopment. However less foresight was shown in the planning of other developments. Potentially indirect flood damage (i.e. caused by disruption and knock-on effects) could be substantial, whether measures to increase resilience to direct flood damages are incorporated or not.

The evolving spatial development strategy for London (Greater London Authority 2001; Thames Gateway London Partnership 2001) envisages that the growth of London will be directed eastwards during the 21st century — a movement that is already well under way. The development of the Thames Gateway area, along the north and south banks of the tidal Thames and Thames estuary is the major component of this strategy. Many of the targeted development sites lie within or close to the flood risk zone (Fig. 2).

Infrastructure which is Highly Vulnerable to Damage and Disruption by Flooding

The vulnerability of London's infrastructure to tidal flooding is very high. Commuters in the exposed eastern part of the city are dependent upon underground and surface rail lines. Underground systems are especially vulnerable to flooding, and congested surface transportation systems are easily disrupted. Because electricity and telecommunication facilities are exposed to possible flooding, they are likely to be vulnerable. However the replacement of copper telecommunication cables with fibre-optic cables is reducing the likely disruption of commercial communications by flood damage. The economic vulnerability of businesses and people in the tidal flood zone is highly variable, with some occupying buildings that are designed to withstand flooding while others, particularly from low-income ethnic minorities, are at much greater risk. With the Barrier in place the risk of tidal flooding is considered to be very low, although of high potential consequence for infrastructure and people within the flood risk zone.

Residual Low Probability Tidal Flood Risks Require Management

Apart from the standard of flood protection afforded by the Barrier system declining as we approach 2030, various exceptional circumstances exist in

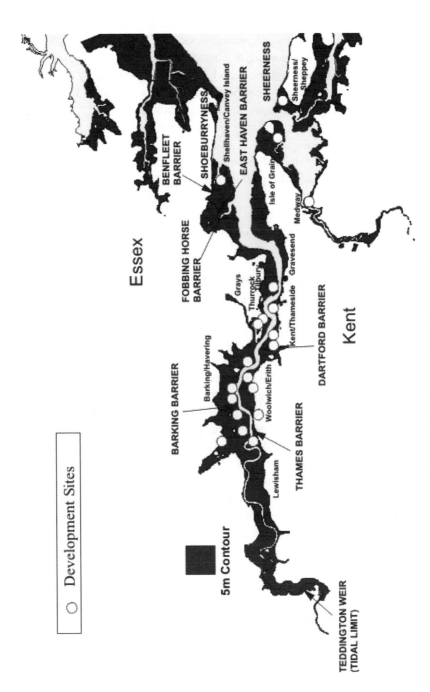

Fig. 2 Thames Gateway: Coincidence of tidal flood risk area and development sites.

which London might still be seriously flooded. Failure to close the Barrier in advance of a major tidal surge through human error in either detecting or forecasting a surge or in operating the Barrier is a remote but theoretical possibility. More likely is failure following a warning to close one of the more than 30 moveable floodgates which make up the entire Barrier defence system downstream of the barrier. Failure to close any of these floodgates could lead to serious tidal flooding in its immediate vicinity, though not to the whole of the flood-prone area. Failure to close openings in flood defences was one of the causes of serious flooding in Northampton in the Easter 1998 floods. The assessed probability of failure to close a floodgate is reported as appreciably higher than the design standard of the Barrier indicating that these floodgates are a weak link within the defence system. Fortunately these risks have been assessed and we understand that recommended risk management actions have been taken. An exceptional tidal surge perhaps exacerbated by vessel impact could cause the collapse of the north bank flood walls just downstream of the Barrier and this could lead to extensive flooding in Newham. Management actions are taken to minimise this risk. Failure to synchronise the closing of the various subsidiary flood barriers along the Thames estuary could produce flooding because closure of the main Barrier generates a reflected wave of water that could surge back downstream. A break-down in the Barrier's closure mechanisms is a further conceivable occurrence, although the risk is very low. The Barrier has ten separate gates. If one jammed, the Barrier would continue to afford considerable flood protection for the relatively limited period of a tidal surge. Also each Barrier gate has four possible power units to operate it.

Rising Tide Levels and Potential Flood Exposure Have Made London Dependent Upon High-technology Investments into the Future

The future physical survival of a large part of London now depends upon high-technology solutions well into the future. The Thames Barrier flood defence system has proved to be a robust defence system to date. However, it involves very costly investment and re-investment in a fixed line of defences that has a limited life-span. Anticipatory approaches are required. Within a few decades at least the structure must be replaced, perhaps by an even larger structure in the outer Thames estuary. As the city's commercial centre migrates eastwards, as the eastern Thames corridor becomes more developed and as the sea level effects of probable global warming become more pronounced, future tidal flood risk problems will be even more difficult to solve than those of today.

MANAGING THE NON-TIDAL FLOOD RISK

Three other types of flooding are common in London: fluvial flooding, thunderstorm flooding, and sewer flooding. Fluvial flooding along the main branch of the Thames is a serious hazard that does not depend on high tides or storm surges. There have been at least six major floods since 1894. They tend to occur after prolonged or heavy precipitation and mainly affect areas along the river from its tidal limit at Teddington westwards towards Reading. Localised flooding of tributaries such as the Lea, Brent, Crane, Wandle and Ravensbourne has been common in the recent past and may be increasing in frequency. In central London, many tributaries were buried or canalised as the city grew and they now must carry increased volumes of runoff. Since the arrival of new housing and industrial estates there has been a large increase in flood peaks in urbanised catchments (Hollis 1986).

In recent decades there have been notable instances of flooding in London as a result of intense rainfall, coupled with surcharging of sewers. Meteorologically these events occur in summer months when clouds 'bubble up' in the moist atmosphere. Such convective storms are usually quite localised with high accumulations of rainfall. The flooding is likely to 'flashy' and quite deep on roads, underpasses and open spaces. There can be a risk to life and property. Weather radar pictures have shown the regular recurrence of extremely intense 'cells of' rainfall that relocalised and may move quite slowly. Examples of severe convective storm flooding in London include the Hampstead Heath flood of 1975 (171 mm in 2.5 hours). Throughout London there has been an increase in sewer overflow floods which have required management action.

The potential impact of intense rainstorm events in London, for example of the kind that occurred in 1998, have almost certainly been under-estimated. Flood defence policies on the tributaries of the Thames in London are quite controversial in some cases, for example in connection with the flood defences constructed along the river Mole and subsequently the Thames at Maidenhead. Given the uncertainty of current predictions about intense rainfall events, severe flooding in London could well occurs in the next few years.

During the past 30 years or so, there has been a major increase in exposure to fluvial flooding in London and surrounding areas, but in reality this is only the extension of a trend which owes its origins to the establishment of London's suburbs throughout the last century at least. To the west of London along the M4 and Thames corridor, from the 1970s onwards, flood plains were developed if they were not part of the green belt. This occurred for example in Datchet, Maidenhead and most Thames valley towns. As the lower part of the Thames estuary corridor to the east of London develops, we can also expect exposure to flooding to increase there as well.

Flood Plains that Continue to Accommodate Development

The typical management response to fluvial flooding throughout London has been a combination of channel clearance, channel maintenance and structural channel improvements. In some cases (e.g. the Lea Valley), flood relief channels were constructed and flood retention reservoirs have been employed (Chapter 12). However, in other cases — for example, the river Thames in the Staines area — the flood hazard has not yet been significantly alleviated owing to unfavourable benefit-cost outcomes of flood alleviation feasibility studies. More latterly flood warning systems have become established, and deployment of Automatic Voice Messaging (AVM) systems has grown as the principal flood warning dissemination system. However, in southern England experience reveals that some people are terminating their use of AVM. In theory, land use controls offer the possibility of preventing further construction in the flood plains (Office of the Deputy Prime Minister 2001), but experience has shown that they are rarely effective because pressures to develop flood plains are intense in the London region. Development controls may slow down flood plain development, but they rarely prevent its accumulation over time. In fact the London region's flood plains are steadily being built upon although some areas are being deployed as natural wildlife corridors (Chapters 5, 9, 11). As developers struggle in the southeast to find locations for a further 4 million or more households, so the flood plains of the region are inexorably becoming developed. Consequently the risk of flooding increasing. This requires the investment in costly flood protection and flood management measures such as flood warning systems. Inevitably these have practical limitations.

STEEPLY-RISING FLOOD DAMAGE POTENTIAL

We have seen that rising tide levels and possible increases in weather extremes, the exposure to flood risk increasing in the London region. But also owing to development of riverine flood plains and tidal flood risk zones, recent research reveals that flood damage potential is rising steeply (Penning-Rowsell *et al.* 2002). This research, commissioned by the Environment Agency and undertaken by Middlesex University Flood Hazard Research Centre, focuses upon assessing the financial and economic losses experienced in the Autumn 2000 floods in England and Wales. However, the research data are nationally applicable and can therefore be applied to London.

The research reveals that the flood damage potential of commercial property has risen by three-fold in real terms between 1987 (Parker *et al.* 1987) and 2000. Technological change is the principal driver behind this steep rise in flood damage potential. Potential residential flood damages have risen by 2.5 fold over the same period, again partly owing to technological change and partly because the growth of new-for-old insurance means that salvage is much less prevalent now compared with the past. The assessment of the costs of the Autumn 2000 floods reveals that the costs of deploying the emergency services were very high and exceeding £5000 per property affected by flooding.

These increases in flood loss potential are quite capable of significantly altering the estimated benefit-cost ratios of possible flood defence projects, such that for example, the projects previously proposed for the Staines area might now generate favourable outcomes. What these results suggest is that decisions about flood defence in the London region need frequent review, and assumptions about what may or may not be feasible need to be challenged frequently.

A RISK ASSESSMENT FOR LONDONERS

We do not know whether London is becoming a more or less hazardous city in which to live and work, yet from time to time — especially when a disaster strikes — this is one of the questions which Londoners ask themselves. One way of assessing London's hazards to determine whether they are increasing over time or whether they are being contained and reduced is to estimate the direction of change in the major factors affecting these hazards. This is not an easy or uncontroversial task but such an approach leads to an overall summary risk assessment for London.

Tables 1 and 2 provide a risk assessment in which a list of hazards affecting London is set against a number of major hazard factors. The overall

Table 1 Estimated risk assessment for London according to major hazard factors, 2002.

Hazard	Risk	Exposure	Impact	Vulnerability of infrastructure	Vulnerability of Londoners	Management response
Sea level rise and tidal flood	H	H	H	H	M-H	H
Fluvial flood	M	M	H	H	M-H	M
Sewer flood	L	M	M	H	M	M
Drought	H	H	H	M	L	L-M
Subsidence	M	M	M	M	L	L-M
Windstorm	M	H	L-H	M-H	L-M	M
Fog	L-M	M	L	L-M	L	M
Snow	L	L-H	M-H	M-H	L	M
Air quality	M-H	M-H	L-H	L	L-H	M-H
Seismicity	L	H	L-H	L-H	M	M
Nuclear radiation	L	H	H	H	H	H
Hazardous wastes	L-M	L-H	L-H	L	M-H	M-H
Hazardous industrial sites	L	M	M-H	M	M-H	H
Transport Crime	L-M	H	H	H	M-H	L-M
Terrorism	M	H	L-H	H	M-H	H
Hybrid i.e. combinations	H	H	H	H	M-H	M

H = estimated as high.
M = estimated as medium.
L = estimated as low.

Table 2 Estimated changes in major hazard factors for major hazards affecting London between 1945 and 2045.

Hazard	Risk		Exposure		Impact		Vulnerability of infrastructure		Vulnerability of Londoners		Management response	
	->50	50->	-> 50	50->	->50	50->	->50	50->	->50	50->	->50	50->
Sea level rise and tidal flood	+	+	+	+	+	+	+	+	–	–	+	X
Fluvial flood	+	+	+	+	+	+	+	+	–	–	+	X
Sewer flood	+	+	+	+	+	+	+	+	–	–	+	X
Drought	X	+	+	+	+	+	X	X	–	–	+	+
Subsidence	X	+	+	+	+	+	X	X	–	–	X	+
Windstorm	X	+	+	+	X	X	+	+	X	–	+	?
Fog	X	X	+	+	+	X	+	X	X	–	+	X
Snow	X	X	+	+	+	X	X	X	X	–	+	X
Air quality	X	?	+	+	X	?	X	X	+	?	+	?
Seismicity	X	X	+	+	+	+	+	+	X	X	+	X
Nuclear radiation	+	?	+	+	+	+	+	+	+	?	+	?
Hazardous wastes	+	+	+	+	+	?	X	X	X	+	+	?
Hazardous industrial sites	+	X	+	+	+	X	X	X	+	X	+	X
Transport	+	X	+	+	+	+	+	+	+	+	+	+
Crime	+	?	+	?	+	?	+	?	+	?	+	?
Hybrid hazards	+	+	+	+	+	+	+	+	X	+	X	+

->50 is the period between 1945 to 1995.
50-> is the period between 1995 and 2045.
+ = estimated to be increasing.
X = estimated to be stable.
– = estimated to be decreasing/worsening.
? = trend is less certain, depends upon socio-economic and political events.

approach is intended to be more significant than the particular ratings which could be improved upon by public survey or 'Delphi' techniques. Nevertheless, the approach leads below to several interesting suggestions. The major hazard factors are **risk** (i.e. the probability of an event such as a flood or transport disaster occurring), **exposure** (i.e. a measure of the number of properties and people at risk), **impact** (i.e. a measure of the likely consequences of an event), **vulnerability of infrastructure** (i.e. the susceptibility of infrastructure to damage and disruption linked to its age and its fitness for purpose including capacity), **vulnerability of people** (i.e. the capacity of individuals and groups to recover because of access to resources and support mechanisms), and **management response** (i.e. the extent to which a strategy exists which effectively minimises risk, exposure, impact and vulnerabilities of infrastructure and people).

In Table 1 a judgmental rating is applied to each hazard according to each hazard factor to determine where the greatest problems are concentrated. In Table 2 an attempt is made to estimate the direction in which the hazard factor is moving during the period of 1945 to 2045 in which we are currently around mid-way in time. The first 50 years of this period is assessed, alongside the second 50 year period. The ratings are suggestions only.

The ratings indicate that London has a range of low risk hazards — a finding much in common with general views about London not being particularly intensely hazard-prone. Exposure to and impact of hazards is relatively high which means that even low probability events are perceived of as potentially very damaging when they occur. Vulnerability of infrastructure is high in comparison to vulnerability of Londoners. This is because Londoners have a relatively high per capita income and are surrounded by support mechanisms including insurance. The rating does not, however, capture the potential vulnerability of relatively poor Londoners and recent migrants because it is too general. As one would expect in a post-industrial society most hazards have been addressed in terms of management strategies but some of these strategies are perceived as being less effective than others (e.g. safety of transportation systems). Table 2 suggests that in contemporary London, risks from most historic environmental hazards are either stable or increasing. Moreover, old hazards are being transformed into new variants. This is clearly demonstrated by the changing salience of different air pollutants and by the tendency of what were previously separate risks (e.g. flood, chemical emergencies) to metamorphose into hybrid or combined ones. The number of low-probability, high consequence hazards has increased.

The most pervasive and uncertain of the changes in natural hazards affecting London stem from climate modifications and the enhanced greenhouse effect. Climatic change is potentially the driving force behind possible changes in many of the natural hazards affecting London including flood, windstorm, drought and subsidence. Recent research has made progress in assessing the impacts of changes in mean climate upon economic and social systems in Britain, although most research on global climate change ignores urban areas. London is likely to be vulnerable to shifts in mean climate and the frequency and magnitude of weather extremes. Because climate modelling is currently insufficiently advanced, considerable uncertainty remains about the links between a slowly changing mean climate and changes in the frequency and intensity of rainfall, winds and windiness. Global warming could precipitate storms nearer to Europe in the Eastern Atlantic and storms might then arrive in Britain in a far more powerful and dangerous state.

Increased exposure to risks is one of the most important trends in contemporary London. This is occurring in part because urbanisation is spreading into previously unoccupied hazard zones, in part because of reinvestment in older hazardous areas, and in part because more and more people are dependent on a few easily disrupted infrastructure systems. It is also occurring because damage potential is rising so steeply. The overall vulnerability of Londoners to environmental risks has declined and probably will continue to do so, mostly because standards of living have increased and are likely to continue to improve. But a widening vulnerability gap has appeared between rich and poor groups during the past 25 years. The increased vulnerability of ageing urban infrastructure is also a major contributor to hazard potential (see Chapter 6).

Management responses to hazards are highly variable. In some cases there have been successful efforts to suppress risks, in others much less has been attempted or achieved. Lessons from the past are slow to be learned — many of the disasters that are occurred in London in the past ten years had similar counterparts in the previous fifty years, and indeed in the 19th century.

HAZARD MANAGEMENT IN LONDON: LIMITATIONS AND OPPORTUNITIES

Hazard Management Response Has Often been Ineffective

Despite a long history of deliberate attempts to reduce environmental hazards, London's responses have often been ineffective. The suggestion is that opportunities for intervention and management have either been lacking or unexploited and that successful loss reduction has occurred by change or as a consequence of circumstances indirectly connected with the hazard. For example, comprehensive land use planning controls have been in existence for more than 50 years, but flood plain development has not been prevented and flood losses continue to rise.

Hazard reduction that has occurred has often been the consequence of fortuitous changes in technologies and lifestyles. The decline of fog and smog hazards after the 1890s probably owed more to dispersal of population and industry in the growing suburbs than to strategies to reduce air pollution by controls on the burning of fossil fuels. Even when smoke control laws were introduced in the 1950s, air quality improvements owed more to improved heating technologies (Elsom 1987). Natural extremes have also failed to generate responsive public policies for other reasons. London authorities made a delayed and confused response to the 1986 nuclear reactor fire in Chernobyl because they judged such an event was improbable and did not warrant planning for. One wonders if a similar mindset affects our approach to potential seismic hazards where in London an event of modest magnitude by global standards might lead to widespread damage. Response has been ineffective for further reasons. The failure to forecast the magnitude of the 1987 windstorm which affected southern Britain including London is an example. In that case failure to invest in off-shore detection instrumentation and to make good use of satellite data led to a gap in the nation's weather warning system (Pearce and Swinnerton-Dyer 1990). The magnitude and severity of the rising floods in Easter 1998 were also not predicted by flood defence staff. This led to a general lack of public warning in the areas affected and to many months of ensuing misery for those flooded out of their homes.

All of these events led to significant improvements in warning procedures, and publication of targets and achievements considerably more open to public scrutiny than those in any other country. The series of rail accidents and disasters affecting London in recent years have other causes, but indicate a lower level of safety management than Londoners have come to expect. Whether the perceived increased risks present in rail travel in and around London reflect another example of hazards being intensified by apparently unrelated policies such as privatisation of the rail network will probably never be determined with any certainty (see Chapter 17; also Evans 2004). This is a clear example of how government policies in one department can spill over with negative consequences into hazard management.

This danger is ever-present and often unappreciated and unaccounted for in the administrative and legal function of government, despite the general desire for 'joined-up' thinking (Fig. 1).

Some Successes Have been Achieved

There are recent examples where deliberate management action — albeit often long delayed and adopted 'just in time' — has successfully prevented a disaster. The experience of the 1953 floods on the east coast of England and the appreciation that the Thames was lapping at the top of the existing flood walls with distressing frequency, created the political will to build a tidal flood barrier. The response was made not a moment too soon as the Barrier was closed soon after it was completed and has been closed many times since. The rapid development of methods for coping with large vehicle bomb threats in the City provides another example of effective response under pressure, and effective action by the authorities may well have thwarted terrorist attacks in London during 2003. On the other hand responses have been less effective when hazards have been perceived to be low-intensity events with difficult-to-identify impacts and diffuse costs. Repeated events that have produced low to moderate losses have often produced ineffective public responses. Flooding has become a widespread hazard in London and much of the problem centres around the city's ageing and deteriorating infrastructure (Fig. 2).

Opportunities for Creative Hazard Management

Debate about hazards in London has often tended to ignore opportunities for integrating hazard management into broader fields of urban policy-making and urban management (Hall 1989; Thornley 1992). Some urban policy interest groups have addressed the topic of global warming, but these often lack executive authority. Even if the main urban decision making entities were cognisant of hazards issues, the scope for improved hazard management would be limited by the lack of a coherent urban management strategy.

Government or Market Directed Policies?

In London opportunities for creative hazard management need to be explored in the context of the major policy paradigms which have dominated Britain's political economy over the past 50 years i.e. government-directed approaches and market-directed approaches. In the hazards sphere these policy paradigms are associated with contrasting, but not necessarily contradictory, values. This contrast is illustrated by the different philosophies. On the one hand there is a view that the state should provide protection for the individual and society and should provide a 'collective memory' for extreme events. The other view is that individuals should be self-reliant and responsible for their own actions and that there is little role for 'society'. Government-directed approaches are sometimes based upon the view that market failures make the market largely inappropriate for delivering effective hazard management. Consequently hazard management is principally a public good, a regulatory approach is necessary and metropolitan government or strong coordinating metropolitan-wide agencies are required for successful hazard management. On the other hand the government also adopts some market-directed approaches concluding that the market can be used to deliver hazard reduction through taxation and subsidy, that deregulation is necessary to encourage enterprise in hazard management, and that a disaggregated approach is desirable rather than metropolitan government. Sometimes a lack of value consistency generates confusion but, as the current Mayor's road pricing for central London initiative illustrates, in heterogeneous, 21st century democratic societies, market-based approaches may be possible only where there is centralised political power. Debates on these broad questions are also beginning in other European countries.

Joined-up Policies for Transport-related Hazards Reduction

London's evolving transportation crises may have a beneficial effect in that they could generate a large potential for hazard reduction. So far public policy has conspicuously failed to address important linkages between transport and land use planning, safety, pollution, health and climatic change. Reducing reliance upon the motor car is the most important goal in addressing these inter-dependent problems. Greater integration of transport systems (e.g. park-and-ride systems), high fuel taxes, road pricing and incentives to transfer car commuting to mass transport systems are the avenues for success. Unfortunately the near collapse of the rail system in 2001 following the Hatfield derailment, and the continuing rail system crises in 2002, have had negative spill-over effects into hazards management and have proved to be massively counter-productive. Institutional change may well be necessary to bring about positive change amongst this nexus of inter-dependent systems.

Early Intervention in Development Planning for More Successful Hazard Management

Early intervention in the development planning process and adoption of an extended time-horizon for planning could make an important contribution to hazard reduction and urban sustainability. Regional planning in England and Wales, is provided by voluntary groupings of London and South-East planning authorities. Single-municipalities are being encouraged by government to influence plan-making at an early stage. The Environment Agency has been devoting resources to catchment management plans which can be used to advise planning authorities about issues relating to water-related risks at the formative stage of plan-making. This may eventually reduce the current rather piecemeal, resource-intensive and often unsuccessful approach to controlling development which exacerbates flood risks. The introduction of additional policy planning guidance for flood risk areas during 2001 may also have a positive impact (Office of the Deputy Prime Minister 2001).

Counter-measures for the Hazard-generating Effects of Urban Containment Policies

The London region provides evidence that urban containment can contribute to exposure of people and property to natural hazards. This is mainly because urban containment programmes generate increased development densities and push development onto hazard-prone land. In the South-East green belt policy may have also contributed to more widespread environmental degradation and to longer-distance commuting with their attendant hazards. This does not mean that urban containment is necessarily bad policy in this context, but because the links between urban containment policies and trends in hazard exposure are poorly recognised, consideration of appropriate hazard mitigation plans and other counter-measures to reduce exposure to risks has been too little. Enhanced design standards for infrastructure and buildings to withstand the forces exerted by hazards, structural protection policies to limit exposure to hazards (principally flooding), improved public information and warning systems, and enhanced natural resource protection policies to preserve and restore natural areas, are all potential counter-measures (Burby *et al.* 2001).

A Londoner's Safety Charter?

Improved hazard management could be encouraged by improving the standards of services that are delivered to the public. Making services, including hazard protection, more responsive to customer needs is one way of approaching this. Publishing target standards, and adopting a philosophy of continuous quality improvement, is a way forward. Many public bodies now publish a 'Citizen's Charter' describing the levels of service that they provide. Currently safety has a low profile within most such charters. Also manipulation of targets and data can easily lead to cynicism. A standards of service approach could, however, deliver a range of hazard management services to the public more effectively by either the public or private sector, and Londoners might well be interested in a Londoner's Safety Charter which integrates approaches to the principal hazards faced by Londoners. Such a Charter might commit to providing Londoners with improved information on a range of risks, warning systems, help-lines and sources of technical advice, and self-protective measures.

New Institutional Designs are Needed

For London and Londoners to respond more effectively to many hazards, mega-city wide strategies and actions are required. This is because many of the natural and social processes that generate increased risk, exposure and vulnerability in London do not stop at Borough or metropolitan boundaries. A mega-city wide institution with strong powers to plan and implement a London-wide transportation strategy, and to link this to land use planning, would clearly assist hazard management. Bottom-up strategies are equally important and urban regeneration initiatives provide a basis for community-based hazard management which addresses a range of hazards such as crime, noise and waste disposal problems.

Neglected Opportunities to Reduce Vulnerability of Londoners to Risks

In the next few decades, urban regeneration programmes could make an important contribution to reducing the vulnerability of Londoners to hazards. Currently in London there is little recognition of the fact that differential vulnerability to risk is an important factor that affects patterns of deprivation and loss. This is partly because data on vulnerability are ill-developed but also because vulnerability reduction is rarely perceived by policy makers as a necessary and practical management response to hazards. Policy-makers focus mostly on reducing risk, and then exposure. But vulnerability reduction rarely gets onto the agenda.

Developing Alternatives to Complement Technological Management Strategies

The opportunities for creating better hazard management will not rely solely on further technological innovation, notwithstanding the success of the largely technological approach taken to dealing with flood hazards in London up to the present time. Technological approaches will always have something to offer in hazard management (as, for example, the current French 'tele-peage' road charging system demonstrates). However our suggestion is that, in maturing its approach to hazard management, London should seek to apply political, social and economic strategies to complement technological ones. This wider approach is entirely compatible with the goal of sustainable development that the UK government has officially embraced (cf Chapter 1).

ACKNOWLEDGEMENTS

The authors are indebted to J. (Ken) Mitchell of Rutgers University for his insightful inputs to an earlier version of this paper.

REFERENCES

Burby, R. J., Nelson, A. C., Parker, D. J. and Handmer, J. W. (2001) Urban containment policy and exposure to natural hazards: is there a connection? *Journal of Environmental Planning and Management* 44(4), 475–490.

Bye, P. and Horner, M. (1998) *Easter 1998 Floods* (2 vols), Independent Review Team, Environment Agency Bristol.

Clout, H. and Wood, P. (eds.) (1986) *London: Problems of Change*, London, Longman.

Doyle, R. (2003) *Flood*, London, Arrow.

Elsom, D. (1987) *Atmospheric Pollution*, Oxford, Blackwell.

Environment Agency (2001) *Lessons Learned from Autumn 2000 Floods*, EA, Bristol.

Evans, A. W. (2004) Inagural lecture Road-safety and rail privatisation in Britain. June 2004; www.imperial.ac.uk/p5315.htm

Gilbert, S. and Horner, R. (1984) *The Thames Barrier*, London, Thomas Telford.

Greater London Authority (2001) *Towards the London Plan*, GLA, London.

Hall, P. (1989) *London 2001*, London, Unwin Hyman.

Hollis, G. E. (1986) Water Management, In Clout, H. and Wood, P. (eds.) London, *Problems of Change*, London, Longman.

Intergovernmental Panel and Climate Change (2001) Synthesis Report, Cambridge University Press, Cambridge.

Mitchell, J. K. (1999) *Crucibles of Hazard: Mega-Cities and Disasters in Transition*, United Nations University Press, Tokyo–New York–Paris.

Nicholls, R. J. (1995) Coastal Megacities and Climate Change. *Geojournal* **37**(3), 369–379.

Office of the Deputy Prime Minister 2001 Planning Policy Guidance Note 25: Development and Flood Risk, ODPM, London.

Parker, D. J. (ed.) *2000 Floods* (2 vols), London, Routledge.

Parker, D. J., Thompson, P. M. and Green, C. G. (1987) *Urban Flood Protection Benefits: A Project Appraisal Guide*. Gower Technical Press, Aldershot.

Parker, D. J., Fordham, M., Portou, J. and Tapsell, S. (1995) *The Flood Risk to London: A Preliminary Scoping Study*, Report for the Metropolitan Police Service, New Scotland Yard, London.

Parker, D. J. (1999) Disaster response in London: A case of learning constrained by history and experience. In Mitchell, J. K. (ed.) *Crucibles of Hazard: Mega-Cities and Disasters in Transition*, United Nations University Press, Tokyo–New York–Paris.

Penning-Rowsell, E. C., Parker, D. J., Crease, D. and Mattison, C. (1983) *Flood Warning Dissemination: An Evaluation of Some Current Practices within the Severn Trent Water Authority Area*. Flood Hazard Research Centre, Middlesex Polytechnic, Enfield.

Penning-Rowsell, E. C., Chatterton, J. B., Wilson, T. and Potter, E. (2002) *Autumn 2000 Floods in England and Wales, Assessment of National Economic and Financial Losses*. Draft Final Report to the Environment Agency, March 2002. Middlesex University Flood Hazard Research Centre, Enfield.

Soja, E. W. (2000) *Postmetropolis: Critical Studies of Cities and Regions*. London, Blackwell.

Thames Gateway London Partnership (2001) *Going East, Thames Gateway: The Future of London and the South East*.

Thornley, A. (1992) *The Crisis of London*, London, Routledge.

Turner, R. K., Kelly, P. M. and Kay, R. C. (1990) *Cities at Risk*, London, BNA International.

Chapter 14

Civilising Transport

David Banister and Elspeth Duxbury

INTRODUCTION

Everyone concerned with the sustainable development of major cities recognises that establishing and promoting the appropriate "civilising" transport policy is essential to the success of the overall goal. This is even more the case in cities than in nations as a whole, because it is in cities where there are greater opportunities and traditions of efficient public transport, and also the political will to ensure its future viability. One of the main reasons is that every economic and social group in cities regards transport congestion as a problem that greatly concerns them, as do public decision makers in many branches of national and local government. The quality of life of Londoners is affected by transport in many ways, both directly and indirectly. London features at number 20 (out of 215) in the William Mercer survey of World Cities (2001). But transport (and crime) scores low on the 39 factors used. In this chapter the future for transport in London is reviewed from a fairly optimistic perspective. Perhaps the recent tendency of environmental degradation to be associated with transport can be reversed, so that transport can instead make a real contribution to improvements in quality of life.

For London as a whole five different quantities are described that indicate the magnitudes of transport demand, use and environmental impact. Three types of area are considered; city centre, local centres and transport hubs. In this chapter, current progress and future plans in improving and civilising transport in relation to these three area-types are made. The five key indicators used in our analysis are:

- Population density
- Mode of travel
- Average trip density
- Air pollution levels
- Pedestrian accidents

and these are now briefly summarised in relation to London.

(a) Population density is one means of showing the variation between the different parts of London, with the black colouring indicating high population densities through to the pale grey at the lower densities in the suburbs (Fig. 1). It is interesting to note that the highest densities in London are much lower than those in other world cities, such as Tokyo and New York, and they also decrease more rapidly over about 10 km, unlike Tokyo where high densities persist for 20–30 km. The empirical evidence from the National Travel Survey suggests that in those areas with higher density people travel less and trip lengths are shorter. Also people there are more likely to use public transport (Banister 2002).

(b) The second indicator is the "mode" of transport defined in terms of the average use over a year of the private vehicle, bus and rail. The value of this indicator also differs greatly from the centre to the periphery. In

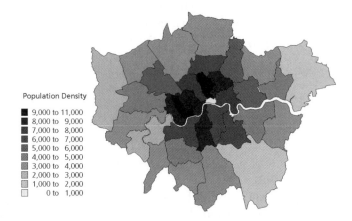

Fig. 1 Population density in London, varying from 10 000 to 1500 people per square kilometre.

the centre, about 85% of travellers use public transport or they walk or cycle. But in the suburban areas about 70% of travel is by car and a further 12% by walking and cycling (Fig. 2). Surveys have shown that in the areas where there is a higher level of car use there are also higher levels of resource consumption and emissions of atmospheric pollutants (Banister 1999).

(c) The third indicator is the average trip density defined as the number of trips originating in a square kilometre, a quantity that also indicates the average trip length (Fig. 3). Apparently, the more journeys people take, the longer they are on average! Here, the highest levels are found in the centre, as this is where most activity takes place, and the levels decline with distance from the centre. The average journey to work time in Central London is now about 60 minutes. But this figure decreases for other parts of Inner London to 43 minutes and to 30 minutes for Outer London. However the local centres outside Central London can also be clearly picked out as areas with a high density of travel trips. There are about 30 such centres across London, each of which mainly serves its local area but some have London-wide connections with longer trip lengths (e.g. Wembley, Croydon and Romford).

(d) The level of air pollution is another key indicator that relates to the impact of transport on people's sense of well-being. Particulates are one

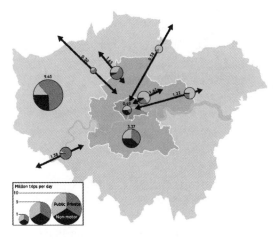

Fig. 2 Mode of travel in London as between bus, rail and private vehicles — the typical journey length is also shown (arrows) and the numbers of trips made by location in different parts of London.

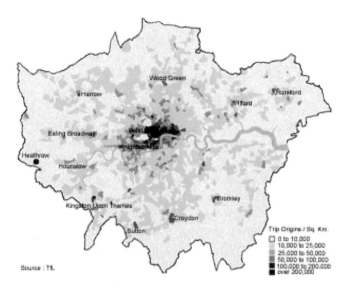

Fig. 3 Trip Density in London in terms of numbers of trips in one year starting within one square kilometre.

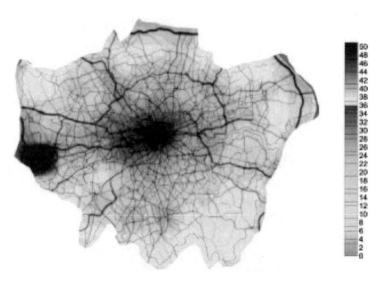

Fig. 4 Levels of very small particulate matter (less than 10 micrometres in diametre that are associated with lung and heart disease) in London (Units: mg/m^3).

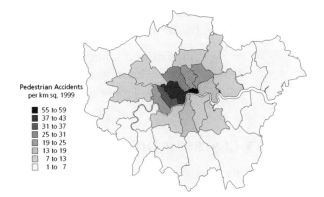

Pedestrian Accidents
per km sq, 1999

■ 55 to 59
■ 37 to 43
■ 31 to 37
■ 25 to 31
□ 19 to 25
□ 13 to 19
□ 7 to 13
□ 1 to 7

Fig. 5 Pedestrian accident levels in London — pedestrian accidents per square kilometre in one year.

of the key pollutants from traffic in London with nearly 70% of fine particulates (those with a diameter of less than 10 microns) resulting directly from road transport. The distribution of this particulate matter in London is shown in Fig. 4, with the locations shown in the darker shading where the air quality was worse than the target standard in 1999. The pollution is shown to be concentrated in the central area of London and to the west around Heathrow airport and along the main arterial roads (see Chapter 7). The relation between levels of air pollution and community health is complex and cannot simply be derived from these figures (see Chapter 6). This indicator is also a measure of noise, which is most severe around Heathrow and arterial roads (Appendix 3).

(e) The most severe indicator of the environmental impact of transport is the average number of pedestrian accidents over a year in a given location. Pedestrian deaths in the UK represent over a quarter of all road deaths and a pedestrian is estimated to be 19 times as likely to be killed in a road accident as a car occupant. London has a rate of 29 pedestrians killed or seriously injured per 100 000 population, and this level is twice as high as that found in the southwest region of England (Fig. 5). In London the highest levels are found in the centre, as this is where the density of pedestrian activity is much greater than elsewhere in London. The accident rates in the eight central London boroughs are 4.6 times as high as those elsewhere in London. Many of these accidents are caused by out of town drivers, since in this area there are low levels of car ownership and car use, with only around 20% of the local population travelling by car in this area.

The measures to be used in civilising transport or making it more sustainable in three different types of location in London will now be outlined — the city centre, existing and new town centres and major international nodes.

LONDON — A CITY CENTRE OF DIVERSITY

In the city centre, all the five indicators outlined above have extreme values, namely:

Density	High
Mode	High levels of public transport and walk or cycle, low levels of car use
Trip Density	High levels and a major generator of trips
Air Pollution	High levels of particulate matter
Accidents	Highest levels of pedestrian accidents

However, within the city centre, there is considerable variation in the character of areas. Some areas are primarily residential and commercial (e.g. the Shoreditch Triangle to the East), others are retail and commercial (e.g. West End), whilst others are regeneration and interchange areas (e.g. Kings Cross in the Northern Edge of the road user charging zone) or business and commercial areas with little housing (e.g. near the Swiss Re building in the City). Although many areas have one main use (such as retailing or entertainment), much of the land use is mixed, giving the opportunity for many work and recreational activities to be carried out in the same area while sustaining and not degrading the local environment.

Action should be directed in five main directions. First, there needs to be a reconsideration of the priority for *the use of the street space.* If the majority of people (around 80%) are travelling by foot, cycle or public transport, then there is an opportunity to provide a greater proportion of the available space to these modes. Streets are primarily for people, with the vehicular movement of people and goods also being fitted in. This dual objective can be achieved if flexible shared spaces are introduced where different uses can complement each other. This has been done in many cities, including London by allocating different uses for particular periods of the day or different days of the week. This allows street markets and parties to coexist with the same spaces being used for deliveries and

Fig. 6 Streets for people in London — this is Monmouth Street in Central London when it was reallocated for use of people for the car free day in 2001.

through traffic at other times (Fig. 6). The purpose is to encourage local street life in residential and other areas, and this in turn may lead to car free zones. As many activities are accessible in the centre of London and as the quality of public transport is improving, there becomes less need for cars to be used for transport within London. In that case, methods are needed to discourage car owners whether living inside or outside Central London from driving in the central area. The UK could do well to follow continental practice where car owners take public transport in city centres and keep their cars for out of town trips, often only taken at weekends.

The second priority should be to create quality *spaces for people* and London has taken a lead here in promoting the World Squares movement with the recent closure of the most scenic part of Trafalgar Square to traffic (Fig. 7). This change again requires a reassessment of priorities with the real-location of space from traffic to people. As with the streets for people, the intention is to create a larger number of quality spaces for people to walk, relax and talk. These are all essential ingredients of the sustainable city.

The third measure is *supporting pedestrian access* through giving people greater priority in locations where traffic is still necessary and high flows prevent easy crossing of streets. The needs of pedestrians should become an integral part of new schemes. It is important to provide information on pedestrian flows and routes, similar to that obtained for other forms of travel. Such information would enable balanced judgements to be made on the siting of crossings, as it will provide information on the total number of people who will be affected by vehicular and pedestrian movements (see Chapter 11).

Fig. 7 Trafalgar Square with the reallocation of space to people — this area had a high density of mixed traffic and high levels of pedestrian accidents until 2003.

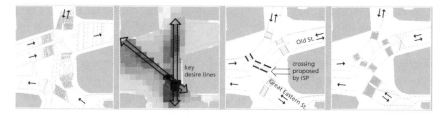

Fig. 8 Supporting pedestrian access — the sequence shows the actual crossing patterns in 2001, the desire lines, the new set of crossing points and the actual.

One example of a new approach is in the Shoreditch Triangle where there is substantial conflict between pedestrian and traffic movement in an area that has a high residential population. Analysis carried out by the Intelligent Space Partnership (Duxbury 2002) identified the current patterns of movement in the areas, and how these related to the crossing of the roads (both at official and unofficial locations). From this data it was possible to calculate the "desire-lines" for pedestrians and, therefore, where it would be most appropriate to locate a new crossing following the redevelopment of the area. Supporting these natural crossing patterns can reduce the levels of pedestrian accidents (Fig. 8). Also increasing the length of the green phase at traffic lights on pedestrian crossings reduces accidents and creates a safer environment for people. This is now the stated policy of Transport for London.

The fourth priority is the *London cycle network*. London has a higher use of the cycle than other major cities in the UK, but a lower level when compared with many other European cities. Policy makers and cycling groups argue that this healthy and efficient form of transport could be more widely used. A London cycle network, which has been talked about for nearly twenty years, is only now becoming a reality. More people will cycle when there are separate networks of routes, as many examples in London boroughs show. The operation of these routes also requires suitable enforcement methods (e.g. to restrict motor vehicles), safe parking and storage for bicycles and better design so that schemes are integrated with other uses of the streets.

The final priority should be the continued enhancement of the *public transport network* within the centre, with the bus being the most flexible and high capacity system. Specially designed "green" routes should be established that give buses exclusive rights of way in the centre over their whole route length. This priority would increase both the capacity of the bus network and the reliability of the system. Such measures have proved highly successful in many other cities in the UK and abroad, and it would further enhance the continued increase in bus patronage in London.

These five priorities in the centre would result in substantial environmental improvement, as road space is transferred from general use into the use to which it is most suited. It also helps to produce a more vibrant city centre, by supporting pedestrian routes along streets or in new parks and squares. Such innovations will help the local retail and cultural economies, as well as creating quality spaces for people. It would mark a dramatic change in priorities away from accommodating as much traffic as possible towards deciding which use has prior claim to particular types of streets and spaces. The balance needs to be redressed from vehicles to people.

LONDON — AS A CITY OF CENTRES

Within London, there are some 30 smaller local high density population and more intensive transport centres, with strong commercial, administrative and cultural activities. Each of them has local and regional functions that complement the function of Central London. In fact London is a City of Centres. In these areas the levels of the five transport indicators are intermediate between low and high.

Density   Low to medium
Mode   High levels of private transport use, low levels of public transport, walk and cycle use

Trip Density   Low to medium
Air Pollution   Medium levels of particulate matter
Accidents   Low rates of pedestrian accidents

These smaller centres need their own particular public transport systems, without which they too will experience the disadvantageous environmental and economic effects of traffic congestion. Because their streets are relatively wide in comparison with those in Central London, it is possible to introduce trams running alongside other street traffic. This may explain the striking economic and *political* success of the Croydon tram, which opened in March 2000 and carried over 16 000 000 passengers after its first year of operation (Fig. 9). This scheme was a design, build, finance and operate concession, with the successful consortium being Tramtrack Croydon Limited (TCL), who have a 99 year concession to run the system, and it is operated as part of an integrated set of services in the Croydon area. There are plans for its extension in six different directions. Trams can link in well with other forms of public transport and they have high levels of capacity. In addition, they can complement the essentially radial network of transport routes in London by providing new circumferential links.

In addition to providing ever more transport solutions to meet growing needs, there are other policies that can be followed that can reduce and ensure the more efficient use of transport. Many of the major producers of traffic (such as businesses, schools, hospitals, government offices and universities) now draw up *travel plans* to suggest the best means by which

Fig. 9 Trams in Croydon (one of the largest of the smaller centres) and buses in Central London.

the amount of car-based travel can be reduced, by the use of measures that encourage car sharing, more use of public transport and the bicycle, and the possibility of home working. This is a process of discussion and involvement, trying to explore the means by which car use in cities can be reduced, and it is seen as a useful means to raise awareness of environmental issues as they relate to transport. Although, many public and private sector employers have cooperated, the main participants have been schools and businesses. Local communities themselves are now thinking about *travel plans* to reduce traffic in their areas. As these plans develop, they can be linked to those for improving air quality being produced by the London Boroughs who are tackling particular pollution "hotspots" where air quality is bad, and in other residential and shopping areas where there are low emissions zones with very low levels of air pollution.

Planning permissions for new buildings can provide clear guidelines on *densities and mixed use development.* In ensuring more sustainable transport policies, through decisions based on encouraging people to live (and work) in higher density locations, there is a possibility that car dependence is reduced provided that public transport is available. As people live closer to their work place, then journey lengths should also be reduced (Banister 1999). These residential densities would be similar to those found in the smaller centres in London such as that at Greenwich

Fig. 10 Greenwich Millennium Village in Docklands.

Village where an architecturally innovative housing development is being constructed on a 32 acre site with accommodation for 1400 flats (Fig. 10). This is a density of about 100 units per hectare, some four times the average for new development in England, but quite comparable with those in Barcelona (see Chapter 2).

These innovative solutions in the "smaller" centres in the Greater London area are partly been driven by the local London Boroughs. There is considerable potential for transport, such as the greater use of trams, greater community involvement, travel plans, and for developments with reduced car ownership or even with no cars at all. Equally important for sustainability are innovations to reduce energy used in housing and other buildings. Here the private sector and non-governmental organisations are taking the lead, stimulated by government guidance and new building regulations. In outer London, the best-known example is the BedZed development located in Beddington, in the Borough of Sutton (www.bedzed.org.uk). Just outside London, Woking has reduced the energy consumption of its council run facilities by 40 per cent over the ten year period to 2001/02.

LONDON — A WORLD GATEWAY

The third category of transport location in London consists of the major hubs for international passenger travel and goods transport. The levels of the five indicators, in these areas have different characteristics to the central and local centres, namely

Density	Low levels of population density
Mode	High levels of private transport use, and low levels of public transport, walk and cycle use
Trip Density	Very high levels — a major generator of trips nationally and internationally
Air Pollution	High levels of particulate matter and nitrogen oxides
Accidents	Low levels of pedestrian accidents

The five international airports in London (Heathrow, Gatwick, Stansted, Luton and City) are the World Gateways, and they handle nearly 120 million passengers each year and an increasing volume of air freight (Fig. 11). Over 80 000 people are directly employed in the airports with many other jobs linked to the airport-related activities (Banister and Berechman 2000). In addition, there are some 7.7 million passengers using Waterloo International station on the Eurostar train. These numbers are

(a)   (b)

Fig. 11 (a) The Heathrow express train that connects Central London to Heathrow in 15 minutes and (b) Planes queuing at Heathrow Airport, often emitting large quantities of air pollution, often for much longer than 15 minutes!

expected to double over the next twenty years (DfT 2002), with the damaging consequences for the environment, human health and the efficiency of all transport systems.

The rate of increase applies not only to the aircraft movements, but also to the means of transport used to access the airport, whether it is by road or by rail (Fig. 11). A major goal for mitigating the effects on the environment is to increase the use of public transport, especially by train, which is highly effective for reducing greenhouse gas emissions and air pollution. At present, about 35% of air passengers and 23% of employees arrive at Heathrow by public transport (2001) and the target is to increase this level to 40% by 2007 (BAA 2002). For Luton, the corresponding figure is 23% of air passengers and some 4% of employees travelling to the airport by public transport (2000). The target here is to increase the air passenger figure to 35% by 2006. But apart from the airport activities, there are many other demands on land close to airports. For example these locations are seen to be desirable by several high tech and other international businesses. Heathrow is the prime example of this, as it forms a second major retail and commercial centre outside of Central London.

All forms of pollution in the Heathrow area are high, with nitrogen dioxide being particularly bad around the airport and the M4 and M25 corridors. This pollutant increases people's susceptibility to lung infections and it has possible links with cancer. The projections from the British Airports Authority (BAA) are not good for 2005 (Fig. 12). If a 3rd runway

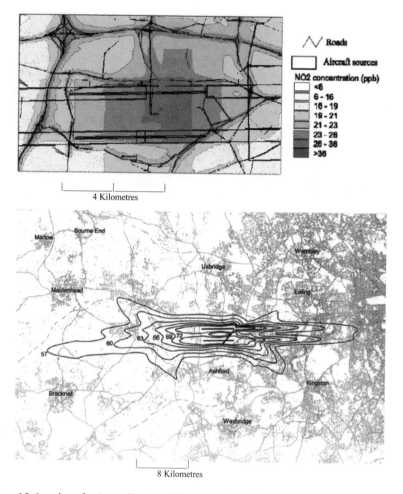

Fig. 12 Levels of air pollution (nitrogen dioxide) and noise levels around Heathrow Airport.

The diagram shows BAA's modelling of projected annual NO_2 concentrations around Heathrow in 2005 with present policies. The pollution at ground level is mainly due to road transport emissions and other sources. The second diagram shows the noise contours around Heathrow, with the outer 57 Leq level (some disturbance) and the central 69 and 72 Leq levels (considerable disturbance). Note that night time disturbance outside these contours, even if only occasional, is also regarded as a serious environmental infringement.

is built here, some 35000 residents may be affected by nitrogen dioxide levels in excess of the EU limits. Even with the introduction of the best available technology, the affected number will still be 5000 residents.

Similarly, the noise contours around Heathrow show substantial disturbance over the whole London area (Fig. 12), with the onset of disturbance occurring at 57 Leq (this is the equivalent continuous sound level or the notional steady noise level over the stated period giving the same energy as the actual intermittent noises, and it is commonly used for aircraft noise measurement) and high disturbance at 67 Leq. There is at present considerable debate over the right of school children to have a quiet environment and of residents to peace and quiet, particularly at night, where even one aircraft landing can disrupt sleep. Measures here are less obvious. Regulations can be introduced to ensure the best available technology is being used and restrictions on the number of night flights have been enforced. But as stated in the recent airports White Paper (DfT 2002), substantial growth in the demand for air travel is forecast over the next 25 years and much of that growth will be in leisure travel, and it will take place at Heathrow.

One approach to dealing with the damaging effects of global warming and air pollution caused by air travel is to raise substantially the prices charged for travelling, so that this revenue can be used to insulate buildings and to reduce engine noise at source. This could mean a substantial increase in airfares. Perhaps aircraft kerosene needs to lose its tax exemptions? Airports could set an example for travel plans, ensuring that at least 50% of all vehicle movements to and from airports are made by public transport. A sustainable transport strategy might suggest a figure nearer 70% as desirable. Consideration should also be given to closing the tunnel at London Heathrow so that only non-polluting forms of public transport have access to the central airport site. It would not be difficult to run a high speed frequent tram link through the tunnel or a fuel-cell bus fleet with exclusive rights of way.

In summary, the growth in air travel and the importance of airports to the economy of London means that difficult decisions will have to be made about whether or not its future growth is sustainable. If the high growth expectations are fulfilled and the economic growth case takes precedence over the environmental issues, then Heathrow (and the other London airports) may pose the largest single threat to London's environment.

CONCLUSIONS

This chapter has summarised some of the ways in which transport in London can be civilised. Much can be achieved through relatively cheap means in the short term, provided community leaders have the vision and the willingness to reallocate priority for space in our streets from vehicles to people and from the car to public transport. As the introduction of congestion charging in Central London (February 2003) has shown, these steps have begun. But there seems to be no reason why progress cannot be accelerated. Major investment is also needed in the transport infrastructure, and much of the Mayor's Transport Strategy is devoted to these high cost improvements in the underground and cross London rail links, as well as new radial bus routes and complementary tram schemes and substantial investment in East London (GLA 2001).

Environmental priorities must be understood as being of equal importance as the economic and social objectives of policy. But environmental policies, if they are to be successful require careful implementation, based on good science and good design, which must match the high standards expected of improvements in environmental quality. There is a need to create spaces within which people want to live and work, and to make them affordable. London is littered with failures in the planning and design of buildings, roads and transport systems. These are partly technical and partly the result of poor communication and consultation (see Chapter 15). In the longer term, this will encourage more people to move back into London, and this in turn will create the possibility for shorter journey lengths and further increases in the use of public transport, walking and cycling. The potential for a sustainable transport system in London grows as the need to travel is reduced.

The alternative is to continue on the same path as now with a continuous growth in congestion and a decline in the quality of life. This alternative is not acceptable and transport must play an increasing role in helping business to achieve greater efficiency. The perceptions of London by investors are also affected by the quality of transport and it may even affect London's competitive position. In addition, the transport system needs to be resilient in the face of natural or man-made disasters that can strike any city, including London (see Chapter 13).

Although we have concentrated on the passenger sector, there are substantial similarities with the freight sector where the same trends are observable. The use of technology both to increase the efficient use of vehicles and to improve vehicle performance is crucial. Goods vehicles make up less than 10% of traffic, but they contribute 30% of nitrogen oxide and 63% of particulate emissions in London (GLA 2001).

Transport policy has in the past been seen as a matter for planners and engineers, and as such has been left to these experts to make decisions about what should be done. This has now changed and it is important to win the support of the communities through their active involvement. Such an approach leads to empowerment and a commitment to change. This support is crucially important when radical alternatives are being considered. The challenge of a sustainable transport system for London is a considerable one, but the ingredients are available for its successful implementation if the momentum that is now building up for radical change continues to grow over the next few years.

REFERENCES

Banister, D. (1999) Planning more to travel less: land use and transport. *Town Planning Review* **70**(3), 313–338.

Banister, D. (2002) *Transport Planning*, London, Spon, Second Edition.

Banister, D. and Berechman, J. (2000) *Transport Investment and Economic Development*, London, University College London Press.

British Airports Authority (2002) *Heathrow Surface Access Strategy*, BAA Review of Progress, May, www.baa.com.

Department for Transport (2002) *The Future Development of Air Transport in the United Kingdom: South East*, A National Consultation, Department for Transport, July.

Duxbury, E. (2002) The Shoreditch Triangle Study, www.intelligentspace.com.

Greater London Authority (2001) *Transport Strategy for London*, Final Version, London, GLA, July, www.london.gov.uk.

Chapter 15

Community Participation in Urban Regeneration Using Internet Technologies

Andy Hudson-Smith*, Steve Evans*,
Michael Batty* and Susan Batty†

INTRODUCTION

The aim of this chapter is to demonstrate how urban planning can be radically changed through communication and consultation by internet and world wide web technologies. These enable communities and individuals to be engaged in designing and regenerating their environment. Such technologies, which involve communicating over the internet using multimedia, are beginning to provide ways in which individuals and groups can interact more effectively and convincingly with planners and politicians in exploring their future. This chapter tells the story of how the residents of one of the most disadvantaged communities in Britain — the Woodberry Down Estate in the London Borough of Hackney — have begun to use an online system which delivers everything from routine service information about their housing to ideas about options for their future. This is located at http://www.casa.ucl.ac.uk/woodberry/ and details of the multimedia, internet

---

*Centre for Advanced Spatial Analysis (CASA), University College London, 1-19 Torrington Pl., London, WC1E 6BT, United Kingdom, asmith@geog.ucl.ac.uk, stephen.evans@ucl.ac.uk, m.batty@ucl.ac.uk.
†Bartlett School of Planning, University College London, 22 Gordon Street, London, WC1H OQB, United Kingdom, susan.batty@ucl.ac.uk.

Geographical Information Systems (GIS), and digital photogrammetric tools we have used can also be seen on other pages of this website.

Because GIS provides a visual medium in which 2- and increasingly 3-dimensional spatial information can be communicated on computers, desktop and also across the web, there is a growing attraction in using GIS for Public Participation. PPGIS, as this movement has come to be called, is a natural consequence of the ubiquity of GIS and our ability to communicate maps and models in an intelligible visual form to those who have no expertise (or interest for that matter) in its technical basis (Kingston 2002). In this chapter, we illustrate how we have developed various aspects of this visualisation in relation to an exciting and progressive example of public participation which is central to the process of urban regeneration in British inner cities (Rydin 1999). Our example, involves a 20 year process of regeneration of social housing in Woodberry Down. This is an area of deprived social housing in the London Borough of Hackney, where participation is now regarded as a permanent part of the process of development. This kind of project will make PPGIS a reality.

We begin with a brief review of new digital media and the way that this is influencing different forms of communication. The Woodberry Down project represents one such form of communication. We first sketch the background to the area, illustrating the critical problems of deprivation which dominate the regeneration that is taking place. We then provide a blow-by-blow account of how the online system we have built, emerged (including the stop-go nature of the funding). There have been many problems in ensuring that the residents and their representatives engage in its use, and in 'wiring' the community to ensure the effective use of the system. We have found that certain technologies are crucial for making sure the system delivers information in a robust and timely manner. Our chapter brings out how this practical project helps in pushing forward the technical frontiers. It also identifies the difficulties in exploiting these new technologies sustainably (Bullard 2000).

THE NEW MEDIA FOR PARTICIPATION

Participation in planning requires information that is strongly dominated by visual media in the form of maps and pictures. Text is an important subset of such data. For online participation, visual interfaces are essential while hardware in the form of computers and networks are critical in such communication. Computers need to be powerful enough to process pictorial information while networks need to have enough capacity to enable users to communicate quickly. This is the real bottleneck in using current systems. Most of those to whom online participation is geared do not have network access other than through traditional phone lines. This limits the speed at which they can receive and transmit visual information. Speed is absolutely critical for systems that deliver media quickly and successfully. Much of our own work in Woodberry Down is focused on developing systems that are workable and robust in these terms. Most systems developed to date are essentially passive, in that information is delivered in one direction only — from the server to the client. Then the only activity on the client's part — the user — is choosing the information that needs to be delivered. Systems where clients acting on such information and feed this back to the server or provider are still quite rare. However, as we shall see, it is essential, if online participation is to move beyond a digital version of simply telling those affected what is planned (Laurini 2001).

Software of course is the key. Good software developed by those who serve information can turn that information into a form that is intuitively digestible by the user, the best examples being good pictures and words that communicate the essence of any issue in the most effective way. Anything more than this depends on setting up communications so that users can act on the information. Technically proficient users might manipulate data for themselves. For example, there are now many websites which deliver numerical data on planning issues which users can store locally, examine and manipulate offline, at their leisure. The London Borough of Wandsworth has a website where those seeking planning applications can examine recent applications and decisions, and map the data. This is a relatively passive form of information delivery for interaction (not selection) between the Borough and the user (http://www.wandsworth.gov.uk/gis/map/mapstart.aspx), although users need to be able to interpret the meaning of such data.

To ensure better understanding and effective participation, maps are insufficient and animated graphics are required. These involve three-dimensional representation of built environments through which users can move and fly. These kinds of virtual reality often require the user to decide what and where to navigate. In this sense they are truly interactive. Web-browsers have long been configured for such navigation but when such information is delivered over standard phone lines, it can be slow and off-putting for the user. The dilemma of course is that the best information about a planning scheme usually requires such representations. Much of our technical work with Woodberry Down involves using and adapting such software so that rapid fly-throughs and related manipulation of visual data is possible over the slowest and least capacity networks. An example of this kind of media is on CASA's *Online Planning* site at http://www.casa.ucl.ac.uk/online.htm. Recently there has been considerable interest in the use of GIS software to encourage participation. In particular desktop GIS has been adapted to web-based processing. Internet map servers deliver information processed on a central server to a client who activates the kinds of function that require some knowledge of the problem. A good example is our *Sustainable Town Centres* site. Users are allowed to take layers of data which indicate various indicators of sustainability, and then to weight and combine these to produce an overall index. The different centres are ranked in order by the degree to which they are sustainable. This online tool really moves beyond the realm of participation, because it is also useful to those who have an expert interest in town centres and have some expertise about the socio-economic functioning of land use and transport in cities (see http://www.casa.ucl.ac.uk/newtowns/ index.html).

Truly interactive participation requires users at each end of the process to act in concert. Bulletin boards, and their graphical equivalent in terms of white boards, act in this way but require active responses and some identity of interest to make the system function. In the Woodberry Down project, the bulletin board cannot be used to its full potential if those who set it up are limited from responding to users because of legal and other restrictions. Slightly shorter and sharper interactive responses between users at the server and client ends have been developed for making group decisions, for networked design studios, and for internet systems that actively involve users in community design. For example, the Architecture Foundation have developed a toolkit for engaging the community in urban design. This has been developed as a passive web resource in which users can follow the design process (see http://www.creativespaces.org.uk). A much more interactive resource — The Glass House — has resulted which enables users to interact with the various design options (see http://www.theglasshouse.org.uk). The Glass House uses state-of-the-art visualisation technologies which can be delivered quickly and at very low cost across the web, a technique we developed in parallel to our Woodberry Down project.

There are many decision-making procedures usually fashioned for experts which involve internet-based communications. These have been developed on local area networks and are gradually being ported to the internet (Jankowski and Nyerges 2001). In general, such decision support systems are not suitable for the kinds of participation we are involved in here. Of greater relevance are totally interactive systems in which there is no assumed hierarchy of users. Chat rooms and related forums are familiar examples of such communications. But where such interactive modes come into their own is through the idea of 'virtual worlds'. Users become more immersed in the online environment, interacting with the media and making decisions about how to manipulate information. Examples of these for design are rare since the notion of developing such visual representations in which users can appear as avatars is highly exploratory. Nevertheless some notable examples show considerable promise for enhanced methods of participation (Schroeder, Huxor and Smith 2001; Smith 2001).

HACKNEY AND WOODBERRY DOWN: DEPRIVATION AND REGENERATION

During the last decade, British local government has been dominated by social, educational and planning problems associated with large areas of public (or social) housing schemes, many of which were finished only a generation ago. The slum clearance programme and the re-housing of a very large proportion of the British population began in earnest in the 1950s. Many inner cities came to be dominated by high rise dwellings under municipal control. These were built to relative poor standards and housed an increasingly deprived population. The run-down in this housing stock was partly caused by poor maintenance. But it has been exacerbated by the migration of the most active and able social groups into owner occupation. In some areas, the public housing stock has been sold off and much of it is now organised into non-profit housing associations. These changes in housing reflect other changes in inner areas now dominated by a series of initiatives associated with regeneration, all of which involve frighteningly complicated sets of policies and instruments (Power 1998). Many of these require the financial underpinning of such actions using variants of the Private Finance Initiative, in which the private sector is encouraged to provide the funds for long term ownership and development of what is essentially public property.

There are 1370 housing estates in England which have been defined as 'deprived' and 112 of these — 8 percent — are located in Hackney which is one of the poorest London boroughs. The best way of illustrating the context is through the index of multiple deprivation (IMD) which is composed of 6 indicators — based on income, employment health, education, housing, and access, with child poverty identified as a critical subset of the income indicator. These 6 indicators are weighted as 25-25-15-15-10-10 and then aggregated to form the overall IMD. When mapped, they provide a picture of the highest relative geographical concentration in the country of social deprivation and sub-standard housing estates. Hackney is one of 33 boroughs in London with a population of around 207 000 in 2001. 40 percent of its population are ethnic minorities and 60 percent of its housing is in the public or ex-public sector. As a municipality, Hackney is not only the second most deprived borough in England but it has the largest concentrations of deprived estates in the land. All 23 of its wards are in the most deprived 10% of all wards in England (where there are 8414 in total). Nine of these are in the top 3 percent and the ward in which the Woodberry Down estates are located is one of these. The pattern of deprivation is shown for Greater London, for Hackney and then for the estates in question in Fig. 1. In fact the various housing blocks that make up Woodberry Down do not contain the most deprived households in the borough, but in terms of the housing indicator within the IMD, this is in the top half of 1 percent of the worst housing conditions in England.

The Woodberry Down estates are in the Woodberry Down and Stamford Hill Single Regeneration Budget (SRB) area, the renewal projects being financed from this source of funds which is bid for competitively each year. In the wards that cover this area, more than 50 percent of all households reside in public housing and if the stock that has been sold off is added to this, then it becomes clear that the area is dominated by estates that are likely to require some substantial regeneration. We do not intend to develop an exhaustive analysis of the demographic profiles of the population for it is clear enough that the populations housed in these areas lack basic amenities. The estates tend to be where the economically and socially worst-off people are allocated housing. There are relatively few long-standing, older residents. There are problems of ageing of course but the key issue is one of poor housing conditions in the first instance. To provide a quick visual impression of the kind of housing that we are dealing with, we show a collage of views around the 25 blocks that make up the estate in Fig. 2. Like so many illustrations, the real sense of how

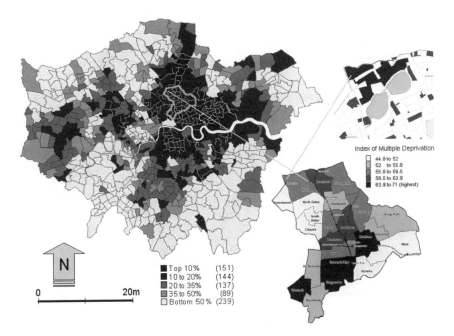

Fig. 1 Deprivation in London, Hackney and the Woodberry Down Estates. The ward in which Woodberry Down is located is in the top 3 percent of the most deprived areas in England and Wales as measured by the 2000 Index of Multiple Deprivation.

run-down the area has become is hard to imagine from these photographs although there is a degree of desolation to the environment which is captured by these pictures.

The area which is to be regenerated is comprised of the estates shown in Fig. 2 which physically cross various administrative and historically integrated, ethnic neighbourhoods. The first housing blocks were developed in the late 1940s by Forshaw as part of his and Abercrombie's vision for London. The forms drew their inspiration from the Bauhaus, even appearing a couple of years ago in the film *Schindler's List*. The oldest blocks are listed as significant historic buildings. There are around 6000 residents in 2500 housing units of which some 29% are owner-occupied. The Woodberry Down Regeneration Team (hereafter called WDRT) have divided the locality into 18 distinct geographical areas although for purposes of resident consultation, these are currently aggregated into 14. There is considerable confusion with respect to tying the official statistics, noted above, to what actually happens on the ground and local surveys

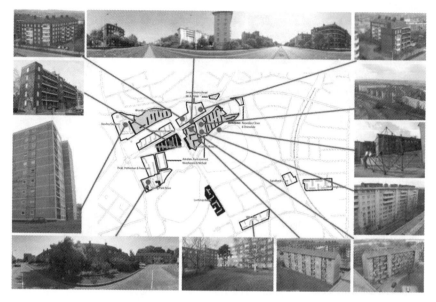

Fig. 2 The estates that make up Woodberry Down.

reveal that in these estates, the white population is in the minority at less than 40 percent with a strong dominance of Black and Turkish populations. These estates permeate the area of Stamford Hill which, incidentally, has the largest concentration of orthodox Jewish population in the UK.

The Woodberry Down project began in 2000 with the establishment of the on-site team. Negotiations began for a Single Regeneration Budget proposal for some £25 m which has since been successful. Currently much of the project is dominated by the negotiation of a Private Finance Initiative to find the major share of the cost which is estimated at some £160 m over 10 years. However, the project did not get off to a good start. The WDRT were located on site in public offices where the local library was situated. The conversion to the team's headquarters and centre led to substantial hostility amongst the local population. The team (WDRT 2001a) reported "Local residents are still angry that not only was their library taken away but also that the centre is, to many of them, not providing any tangible benefit or service to the estate. The WDRT believes that this is not because of the fault of the resident managers but due to the conception and delivery of this project" (page 11). In fact what this issue revealed was that there was already substantial community participation and representation in the area on which the entire project was attempting to drawn in managing the regeneration.

There are now nine Tenants' and Residents' Associations in the area with another two in the process of registering. There are six estate committees serviced by Hackney Council and these meet quarterly. The Stamford Hill Neighbourhood Committee meets nine times a year and is attended by Council officers and local Councillors. The Council's housing stock in Stamford Hill is managed by the Paddington Churches Housing Association and there is a monthly tenants' panel who discuss management. The Estates Development Committee (EDC) which has been set up to represent the regeneration of the estate cuts across these. It currently has 27 members whose role is to liaise with the WDRT and to represent the views of those affected by the considerable disruption that is about to occur as the regeneration gets under way. The process of online participation has been both motivated and endorsed by the EDC and the WDRT, and the website reflects the close involvement of this Committee.

The WDRT have spelt out on behalf of the Council and the community very laudable and ambitious aims for the project (WDRT 2001b). These involve conscious 'bottom-up' consultation and involvement at all levels. In the first year, 12 community meetings were held, while all households had been contacted and the various meetings had formally involved over 20 percent of all residents in the area. The vision shared by everyone in the project is that the estate should be developed sustainably to the benefit of the community and the local ecology. The project has the explicit goal of utilising new technologies to deliver services and involve people more effectively. All these are consistent with the UK government's 'modernising Britain' campaign. The worrisome aspect of the project, like most such initiatives in Britain at present, is that it is continually being monitored while at the same time requires efforts to seek further funding. These activities continually threaten the progress of the scheme and divert valuable resources to open-ended and inconclusive debates about showing 'best value for money'. We are currently three years into the project. Twenty-five million pounds has been committed, while £135 m still has to be raised. Designs have still to be prepared, and there is little to show on the ground so far. We believe that the web resources we have developed at least go a little way to pushing what is clearly a tortuous process forward, and to these we now turn.

DEVELOPMENT OF THE WEB RESOURCES

The decision to develop an online method for participation emerged in early 2000 from a series of related projects that involved projects in Hackney. The catalyst in many ways was the Hackney Building Exploratory, a community-based initiative which enables local communities to learn about their local environment and to participate in ideas about making it more liveable. The Exploratory is located in an old school within the borough and is full of fascinating models and maps of the community, built professionally from standard materials as well as informally by children and adults as part of their educational visits. In 1999, we began to develop a series of digital exhibits which complemented the material exhibits. This has enabled local residents to examine planning information using the latest ways of visualising development by gaining access to this media across the internet. It led to the direct development of educational software which let visitors to the Exploratory find out about the local community using GIS, digital panoramas of street scenes within Hackney, different types of housing within the borough, and patterns of deprivation and disadvantage within the East End of London (Batty and Smith 2001). Computers were located in the Exploratory and were an instant 'hit' with children who form a very large proportion of visitors. A web-based version is available at http://www.casa.ucl.ac.uk/hackney/.

The Exploratory was also involved with the Architecture Foundation, a charitable trust devoted to promulgating good architectural design which has a strong community influence and is supported by the leading architects and planners in the UK (http://www.architecturefoundation.org.uk/). It also had good contacts with Hackney borough whose GIS team were actively seeking ways of extending the relevance of their work through other digital media such as 3D visualisations. Moreover the Architecture Foundation were organising the British entries to *Europan 6*, a competition for young architects. From these entries one based on a Woodberry Down site was chosen. In mid-2000, the Architecture Foundation and the Exploratory also began to explore funding for a wider London-based project involving online community design and we were involved in proposing various extensions to projects that we had already developed as a basis for this. The development of online web-based resources for public participation in Woodberry Down emerged as part of these proposals. The crucial issue was the development of multimedia content in a sufficiently intelligible form for residents to use in thinking about future design options for the community.

The WDRT decided to fund the project in late 2000 after they were convinced that we had multimedia methods fast enough to deliver visual content to the site. We had been perfecting these techniques using an area of central London around the BT Tower, adjacent to our offices, and in an application for British Nuclear Fuels at Dounreay where the reactor was being mothballed. The Architecture Foundation acted as brokers through which the project could be run using their charitable status and in early 2001, the WDRT laid out preliminary ideas for the structure of the website. A rough draft of the site was made and we then began to meet with residents to illustrate what might be done and to test the extent to which the media that we were proposing was acceptable to individual users where those concerned had only the most basic of IT skills.

The structure of the site is divided into four different areas. First there is information, mainly in the form of text about the process of regeneration which occupies at least half the site. Data in the form of reports can be downloaded from this area but the main focus is on explaining what is happening in terms of the regeneration process and informing the residents as to the situation with respect to their own housing. The second area is mappable information supported by panoramas which is currently quite exploratory in intent as eventually this will be used to allow residents to get some feel about what the future of the area might be like. This material makes use of fast multimedia and currently portrays the area as it stands. This is particularly useful for those who are not familiar with the area. The display illustrates how it is possible to visualise the physical aspects of the area, even when communicated over low capacity bandwidths. The other two parts of the site are much more interactive. The third part is a bulletin board of fairly standard form which enables anyone who is registered to post comments. The fourth is the most experimental. Currently this shows how different designs for the future can be viewed in 3D, and also how the building and landscape designs could be changed and moved around the site. Four options are currently present. So far, these involve 300 m, pan and the capability move elements around. Once the system has been further developed, it will allow residents to engage in their own designs and post their schemes to the WDRT and other groups with an interest in the future of the estate.

The first proposal was to use an internet map-server to deliver maps online which residents could query. However, it is not really possible to use typical map servers for the kind of purpose we have in mind here. Residents do not want to query a map, but they do need to see visual information in 2D and 3D very quickly. They need to be able to do this over standard telephone lines. Thus although ESRI (UK) donated a copy of *ArcIMS* map server for this purpose, It was necessary to move to much faster and simpler media, developing and using freeware/shareware based on various software products developed by *Viewpoint* (http://www.viewpoint.com/). In fact, the joint development of the website and the testing of different media in hands-on form with the Estates Development Committee (EDC), were presented at some of the highly contentious residents' meetings. These meetings mainly dealt with community issues involving the process of regeneration, but they were also used to demonstrate the website.

A particularly innovative feature of this project was the decision to engage the resident representatives in the EDC directly in the design of the website. As part of the overall funding, monies were set aside to purchase enough computers and internet access to put each representative online with the computer and its access located in their own dwelling. But it was agreed that representatives would use their access to engage their wider community in the participation process. This decision led to many difficulties. The notion of a public authority providing residents with free computers; the fact that their usage could not be controlled; the requirement that representatives would engage those who they represented in their own homes — all these were highly controversial and debatable issues. The notion too that if representatives did not use their computer, they would be taken from them also raised difficult issues. As a result, the computers once purchased remained in a warehouse for 6 months before the Council agreed to their release. There was a danger that homes would be wired and then demolished or refurbished. Clearly this is a far reaching issue — that to replace physical infrastructure one may need to add to that infrastructure before the replacement takes place.

The data base was constructed over a four week period in the early Spring of 2001. A massive number of panoramas were photographed at roof and ground level and these form the various visual sequences that are embedded in the website. *Zoomview* and related products from Viewpoint, were used for fast animation, zooming and panning of the aerial photographic coverage of the site which is used as the basic locational referent. Essentially these products generate views using a data-streaming technique called 'pixels on demand' in which a scene is divided into a large number of small pieces, each piece delivered being dependent on the pan and zoom within the given window that is selected. The scene is quickly refreshed to produce the greatest detail but the user has a clear idea of what the overall scene looks like while this process is going on. The *Viewpoint Media Player* (*VMP*) is required for this but this is now common on many machines and comes 'bundled', for example, with standard PC software. It can be downloaded over a standard phone line in a couple of minutes and the request to do so is always activated when a Viewpoint scene is generated. The software enables IT designers to layer information too and to link the scene to other web-based software such as *Flash*. We decided very early on that *VRML* would generate 3D file sizes far in excess of what might be handled by a basic user and thus the focus of software on the site is no more elaborate than the fast graphics that can be read by *VMP*. In fact, to develop the site we had to collaborate with Viewpoint who were quite literally writing elements of their software while we were using it.

The initial website design was meant to run until April 2001 but because the graphics design team who produced a first draft of the site and the Woodberry Down logo were slow to start, a working prototype was not available until the summer. Four versions were developed during these months and this represented a detailed collaboration with WDRT and the Architecture Foundation over many issues. Moreover, involving the residents was painful at times. For example, although the EDC are central to the design of the website, representatives wanted to remain anonymous with respect to their locations/addresses in case they were identified as those having free computers and their homes then burgled. This might seem fanciful to many readers but in east London, such fears are quite normal. The website was finally launched in November 2001 along with the exhibition of designs submitted as part of the *Europan 6* competition.

THE STRUCTURE OF ONLINE PARTICIPATION IN WOODBERRY DOWN

As we mentioned earlier, many online resources for participation are only one way; that is interaction by users is passive, being based on rarely anything more than email and comment forms. However in Woodberry Down, interactivity — two way communication between providers and users as well as between users themselves — is central to the process and the website is thus configured to contain various comment forms, bulletin boards, animations, fly-throughs and pictorial manipulations. As the website is continually under development and will evolve with the process of participation and the schedule of regeneration, we will soon add sketching facilities, as well as policy forums for online debate. The structure we have designed is strongly orientated towards low level but comprehensive interaction, is geared to online discussion, and has a clear focus on community design. Professional experts and the community are the target users and providers although political representatives are also likely to feature in its use.

The website has a particularly simple organisation. Essentially there are four main types of information: *textual information* about the entire process of regeneration and the site itself, services, and related facilities; *multimedia* as maps and panoramas about the various component housing blocks which make up the estates; *design options* reflecting the kinds of designs that might be developed for the site; and a *discussion forum* which enables users to interact with the WDRT concerning any aspect of the regeneration process. Textual data forms the vast majority of information that the site is able to deliver and this is accessed as pages through various drop down menus accessible from the home page. These menus cover seven topics: What We Are Planning, 3D Virtual Tour, Regeneration and You, Your EDC, Background and Research, Community and Services, and Youth and Kids. We show a version of the home page in Fig. 3 which

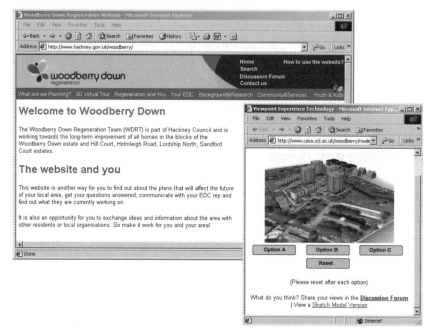

Fig. 3 The Woodberry Down website with inset window showing the Viewpoint Media for exploring various housing options.

contains the design option which we will examine below, as an inset, showing how any user might access the home page of the site while also opening up other windows from the site itself.

What We Are Planning gives access to four pages — relating to the vision for the future, the partnership, yet to be chosen, which will enable the site to be developed through various private finance initiatives, the first stage of the works with access to the 'decant status' of the various housing blocks, and the planning brief. The process of regeneration is plagued by esoteric terminology and acronyms and under the menu associated with **Regeneration and You**, there is a section on 'frequently asked questions' (with answers), and a 'jargon buster' which defines the various terms used by officials such as 'Basic Credit Approval'. There are links to the tenants' temporary rehousing or 'decant' status page and to housing advice — links to other housing agencies from associated pages, while under **Community and Services**, there are links to housing management advice and local services, all of which lead to their own pages. There is a section here that lets users provide the WDRT with information about local events. **Background and Research** provides a brief history of the area as well as key documents referred to as 'Yellow Books' about the regeneration which can be downloaded as *Acrobat* PDF files. This illustrates the sorts of problems that we have had to grapple with. PDF readers are free whereas documents which are set up in *Word* files require the appropriate software, which is not free. Yet PDF is a much less intelligible format for the average user.

There is extensive information about the EDC accessible from **Your EDC** menu which gives information about the constitution of the committee, how often it meets, what it does, and its local representation. Pages dealing with **Youth and Kids** are under construction and currently simply display graffiti and such like in the environment. As the site is under active development, visual information about the existing site and future plans are contained under the **3D Virtual Tour** menu which lets users select from 104 blocks, load pannable and zoomable aerial photographic maps, and thence select digital panoramas of different parts of the site giving some feel for what the place is like now. These use the Viewpoint media introduced in the previous section. If the user zooms into an area of the map, then a panorama is loaded and using a sequence of point and click, this panorama can be opened up from a spherical window and the user can get some sense of the physical conditions and space of the housing in that area. Currently this facility is, as implied, a 'tour' in that it simply illustrates what is possible but in time we intend this to be integrated into the sketch planning capability which we are developing in another area of the site. In Fig. 4, we show typical examples of the visual panoramas and zoomable map layers that can be accessed for all housing blocks on the site.

The area of the site where we are developing the sketch planning capability which will go online once the design options stage is underway, is currently accessible under the Wired Communities menu item which appears in the drop down menu **Background and Research**. So far, we have only developed typical options for Rowley Gardens; there we present three options which enable the user to see the present configuration of housing blocks and to test three alternative designs which can be explored in 2D and 3D. The initial screen shows the existing housing which is composed of a mix of high rise blocks and low rise. The three options when activated replace the existing buildings thus giving the user a sense of how the estate would look. We need to do much more to make this effective but the tools are being developed and we are encouraged by the fact that residents are excited by these possibilities. We show the existing housing and three options in Fig. 5.

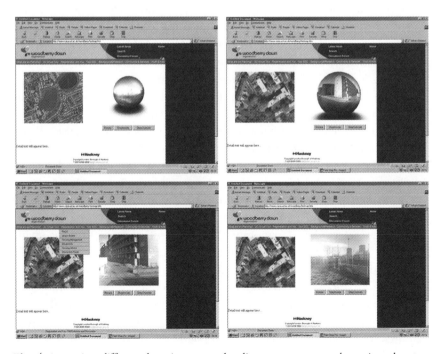

Fig. 4 Accessing different housing areas, loading panoramas and moving about.

(a)   (b)

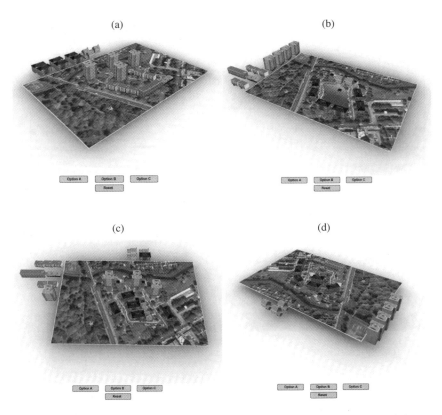

(c)   (d)

Fig. 5 Options for the redevelopment of Rowley Gardens. When a user clicks on one of the options, the current configuration of housing at (a) above moves to the side of the map and new housing options automatically assemble themselves in (b), (c) and (d).

The last feature we will note involves the Bulletin Board, or Discussion Forum. This feature went online in March 2002 and immediately residents began to post notes to the board. One of us (AH-S) acts as the moderator and manages the site but once the material was posted, then the WDRT needed immediate involvement. Many of the postings relate to services to be provided by the Council as part of their role as landlords of public housing since many of the messages are critical, it was resolved that, for legal reasons, the WDRT is barred from responding for fear of litigation. This is a major obstacle to the very notion of participation and it shows no sign of being resolved. It further reinforces the general feeling amongst many residents that local government is hostile, remote, uncaring and even

Fig. 6 The discussion forum with a 'Typical' message.

malicious and this does not bode well for the process. To give some idea of the issues, in Fig. 6, we reproduce a typical message from a user to give an idea of the power of discussion as well as the nature of the argument.

CONCLUSIONS: WHAT NEXT?

One of the key issues emerging from this chapter is the need to develop participation which is truly interactive. This should be genuine two-way dialogue where providers of information respond to users, and where users respond to providers in an ongoing collaborative process. Establishing such a system is easier said than done. Most participation schemes tend to be short-lived and somewhat passive. Information is provided and sometimes meetings may be held to consult those affected, but any dialogue is often too late for any effective action by the community. In online participation, by contrast, especially where this involves providing mapping to the community — in PPGIS for example — the interaction is more effective, but the challenge for everyone involved is even greater, as the whole approach and technology is less familiar. The nature of this media is such that there needs to be active use and provision. Often participation schemes are financed as one-off ventures. But in the case of web-based dissemination it is essential to provide sufficient funding for ongoing development and maintenance of the media.

Any website needs to be continually maintained and updated as it is a portal for the dissemination and communication of continually changing information. It might even be argued that if the aim is to provide those affected with information where none has been provided before, then there are traditional methods of information that are better than the web, such as distributing leaflets through doors. Where there is the need to involve every house in the community in detailed consultation over a period of time, then these methods reach more people than the web. Web-based consultation, even if it is selective, is still attractive in that it can incorporate responses quickly and can chart and communicate changing circumstances. But its limits must be recognised. In the Woodberry Down project, the single biggest challenge is in convincing the WDRT and the community of the need for continuing involvement. It is difficult to fund this kind of participation in competition with the wider portfolio of participatory activities. The very existence of the regeneration team in the local community is certainly enabling the website to be publicised, but the project is currently maintained and updated by ourselves as a labour of love.

There is every prospect that online communication and consultation will grow as more and more residents acquire computers. And there will be a growing consciousness of similar projects around the world as the internet continues to pervade our daily lives. In their mission statement, the WDRT believes "… that it needs to build a trusting working relationship. A real partnership, with residents. It is certain that when it comes to involving the local community in the regeneration, the quality of their involvement in the process may well be more important than the final outcome of many key decisions" (WDRT 2001a, page 18). Online participation is showing that it is an effective means to achieve this involvement.

REFERENCES

Batty, M. and Smith, A. (2001) Virtuality and cities: definitions, geographies, designs. In Fisher, P. and Unwin, D. (eds.). *Virtual Reality in Geography*, London, Taylor and Francis, pp. 270–291.

Bullard, J. (2000) Sustaining technologies? Agenda 21 and UK local authorities' use of the world wide web. *Local Environment* **5**, 329–341.

Jankowski, P. and Nyerges, T. (2001) *Geographic Information Systems for Group Decision Making: Towards a Participatory Geographic Information Science*, London, Taylor and Francis.

Kingston, R. (2002) Web-based PPGIS in the United Kingdom. In Craig, W. J., Harris, T. M. and Weiner, D. (eds.). *Community Participation and Geographic Information Systems*, London, Taylor and Francis, pp. 101–112.

Laurini, R. (2001) *Information Systems for Urban Planning: A Hypermedia Cooperative Approach*, London, Taylor and Francis.

Power, A. (1998) *Estates on the Edge*, London, Palgrave Macmillan.

Rydin, Y. (1999) Public participation in planning. In Cullingworth, J. B. (ed.). *British Planning: 50 Years of Urban and Regional Policy*, London, Athlone Press, pp. 184–197.

Schroeder, R., Huxor, A. and Smith, A. (2001) Activeworlds: geography and social interaction in virtual reality. *Futures* **33**, 569–587.

Smith, A. H. (2001) 30 days in Activeworlds — community, design and terrorism in a virtual world. In Schroeder, R. (ed.). *The Social Life of Avatars: Presence and Interaction in Shared Virtual Environments*, Berlin, Springer-Verlag, pp. 77–89.

WDRT (2001a) *Community leadership at Woodberry Down*, Woodberry Down Regeneration Team (http://www.hackney.gov.uk/woodberry/pdf/community.pdf).

WDRT (2001b) *Vision, Objectives and Procurement* (2nd Edition), Woodberry Down Regeneration Team (http://www.hackney.gov.uk/woodberry/pdf/vision.pdf).

Chapter 16

Towards a Sustainable Environment for London

Lorna Walker and Matthew Simmons

INTRODUCTION

"at their best, cities are the most beautiful and joyful manifestation of civilised life".

Lord Rogers of Riverside

Cities are the social and economic drivers of countries and whole continents. Their institutions provide the focus of culture, architecture, art and science. If civilisation is 'an advanced stage of social development (Anon 1964), our major cities should be the most civilised places in the world, yet we only have to look to some of our larger conurbations to realise that 'civilisation is only skin-deep; scratch it and savagery bleeds out' (Fernández-Armesto 2000). With 90% of population growth in developing countries occurring in urban space (WECD 1987), the pressure is on to ensure that cities become more sustainable.

Both Government and industry are beginning to realise that our current systems of development can only provide for a small section of global society, and that a change must be made towards a more sustainable environment. Governments have taken on new policies and public spending programs. Industry has identified sustainability as a goal as well as financial and social objectives. Consumer power has emphasised the importance of environmental factors in production. These are all phenomena of the last ten years. They signify a change in thinking across society as a whole. World cities as organisms must respond to and be drivers of these changes, as they provide for the needs of the people who live and work there, now and for generations to come.

Sustainable development and urban regeneration are complementary; one will not work without the other. In this chapter some of these new policies and ways of thinking will be outlined. Through careful planning and continued monitoring and assessment, London can continue to be a world class city, providing a good quality of life for its citizens and the people that rely upon it for their livelihood.

SUSTAINABILITY IN BRITAIN

The Rio Earth Summit in 1992 called for Governments to formulate and publish a strategy for sustainable development within their countries. Britain was one of the first to publish a strategy in 1994. The new Labour Government in 1997 confirmed its commitment to sustainable development. It published in *A Better Quality of Life* (HMG 1999) a sustainable development strategy for the UK, and in *Quality of Life Counts* (HMG 1999), its principles and indicators of sustainable development. These included benchmarks against which progress could be measured. The strategy has received three annual reviews from the Government's Sustainable Development Commission, each entitled Achieving a Better Quality of Life (DEFRA 2003).

At the same time, the Government was addressing the problem of declining urban environments with the formation of the UK Urban Task Force, set up by John Prescott MP and chaired by Lord Rogers of Riverside. In 1999 its report *Towards an Urban Renaissance* was published, aimed at providing guidelines for the Government to achieve urban regeneration. The five key elements identified in the report were:

- The sustainable city
- Making towns and cities work
- Making the most of urban assets
- Making the investment
- Sustaining the renaissance

The Urban White Paper published in October 2000 contained 105 of the recommendations made in the Urban Renaissance report. The Urban Summit recommended by the Task Force to happen within two years of publication took place in November 2002, at the NEC in Birmingham. Over 1600 delegates attended the Summit, including a number of Ministers, and speakers including Prime Minister Blair, Chancellor Brown, and Deputy Prime Minister Prescott. Significantly, Gordon Brown spoke of £1.9 billion for Neighbourhood renewal over this parliament, and John Prescott of the Sustainable Communities Plan (formally launched February 2003, this is a £22 billion plan to enable Government and stakeholders to co-ordinate efforts to bring about sustainable development). Prime Minister Blair spoke of a five year strategy to improve parks and open spaces, and launched the anti-social agenda to deal with, among other things, vandals of public spaces.

PRINCIPLES OF SUSTAINABILITY

"Our biggest challenge in this new century is to take an idea that seems abstract, sustainable development, and turn it into a reality for all the world's people"

Kofi Annan

What 'sustainability' means to those responsible for providing 'sustainable development' has been debated for some time. Perhaps the most familiar definition of sustainable development — *"development that meets the needs of the present without compromising the ability of future generations to meet their own needs"* — was provided in 1987 by the Brundtland Commission. Herman Daly, Senior Economist at the Environment Department of the World Bank for six years, gave a definition of sustainability as the capacity of the ecosystem to sustain flows while keeping natural capital intact (Herman 2002).

These definitions encapsulate the work at the United Nations Conference on the Human Environment (Stockholm 1972) which produced the principles that form the basis of Agenda 21, the action plan adopted at the United Nations 'Earth Summit' on Environment and Development in Rio 1992.

The *Bellagio Principles*, intended to guide practical action, were determined in 1996 by an "international group of measurement practitioners and researchers" brought together by the International Institute of Sustainable Development with support from the Rockefeller Foundation. The following principles were unanimously endorsed:

- Guiding vision and goals
- Essential elements
- Practical focus
- Effective communication
- Ongoing assessment

- Holistic perspective
- Adequate scope
- Openness
- Broad participation
- Institutional capacity

Principles such as these must be included in all projects; sustainability can only be achieved through their systematic implementation. They are also used as guidance to develop useful and holistic *indicators* of sustainability, which allow the tangible measurement of a project's contribution to sustainability.

The UK Government prepared objectives and principles of sustainable development, applicable throughout the UK which have been tailored by many regions to meet their specific needs. The DETR publication *A Better Quality of Life* suggests the following goals are essential considerations for any project:

- Social progress that recognises the needs of everyone
- Effective protection of the environment
- Prudent use of natural resources
- Maintenance of high and stable levels of economic growth and employment

And *Achieving A Better Quality of Life* suggests the following principles and approaches to help meet those goals:

- Putting people at the centre
- Taking account of costs and benefits
- Combating poverty and social exclusion
- The precautionary principle (which should include considering resilience against natural or man-made climates, Chapter 13)
- Transparency, information, participation and access to justice
- Taking a long term perspective
- Creating an open and supportive economic system
- Respecting environmental limits
- Using scientific knowledge
- Making the polluter pay

A formal method is required to apply these principles and objectives to actual projects; there should therefore be a framework to assess sustainability.

SUSTAINABILITY ASSESSMENT

"an easily observed variable that may be measured at low cost and is highly correlated with the state of a complex system of interest for decision making"
Douglas Pachico, on indicators, 1996

For a system to be sustainable a number of issues, which tend to have complex interconnection must be accounted for. Sustainability itself is a concept which, when applied to such a system, is essentially a goal for future achievement. Therefore we have to ask whether an imperfect system is close to approaching the goal of sustainability, and over what period will it reach this goal if a sustainable strategy is adopted. We need some ways of assessing where a system is, and analysing the progress that the system has made. The United Nations, and in the UK the DETR (now DEFRA), recognised this need and have defined indicators for use in assessing sustainability.

Establishing indicators implies that those involved want to assess and understand how they are reaching various ideals. Indicators create interest and value in what we measure, for example, effectiveness. Some values are culture-specific, but since the sustainability principles discussed earlier are accepted in most societies, they can be applied world-wide when developing indicators. In *Quality of Life Counts* the UK Government proposed 15 headline indicators giving a broad view of sustainability, and a further 150 indicators focusing on specific issues to identify areas for action. Indicators include:

- Equal Opportunities
- Freight Traffic
- Crime
- Environmental Management Systems
- Emissions
- Green Space
- Security
- Key Services

Each indicator has supporting data that needs to be collated and analysed in order to appraise its performance. Data can include employment and education policies, scientific measurements such as air quality data, levels of crime and availability of services.

The choice of indicators is a critical determinant of the behaviour of the system. They provide evidence of how well the system is designed and analysed, and influence the people involved and their overall behaviour.

Although working within a framework defined by indicators does not guarantee results, results are impossible without proper indicators which, in themselves, can encourage looking at the system from a multitude of angles. A model is necessary to apply them in a practical and functional way. Models used in sustainability include:

*   "Pressure-state-response" models — used to 'measure' national sustainability
*   System-based models — focus on interactions of systems (e.g. Ecological Footprint model, Chapter 5)
*   Indices — ranks a country's performance on human development criteria (e.g. UN Index of Human Development)
*   Thematic models — used to assist in decision-support (e.g. SPeAR®, see below)

One such model is the SPeAR® sustainability appraisal tool, developed to provide a framework for assessment and also a graphical representation of sustainability. The model has been used successfully in a number of applications, such as:

*   Comparison of sites, processes or products
*   Assessing designs, policies and strategies
*   Tracking sustainability of a project (e.g. life-cycle performance)
*   Aiding information management — decision making
*   Providing auditable information
*   Highlighting corporate sustainability goals
*   Incorporating best practice
*   Showing transparency of approach
*   Providing opportunity for innovation
*   Monitoring different elements sustainable development

SPeAR® is an internet-based software program that uses a scoring system to assess a number of indicators based on those recommended by the DETR (now DEFRA) and checked against UN and EU guidance. There are some 150 indicators in the program, 119 of which are *core* indicators which to ensure consistency cannot be changed although an initial review process should consider whether the indicators cover all the relevant factors for the particular study, and indicators may be added to ensure the appraisal addresses all issues.

Once the indicators are finalised, the data to be appraised must be collected. This might include documentation review, site walkovers, and meetings with stakeholders. Workshops may also be used to discuss the indicators and obtain views from interested parties. Once the information is gathered, the appraisal can begin.

The appraisal scoring system uses a seven-point scale to rank the system's performance on each indicator:

- +3 = best case scenario
- +1 = best practice
- 0 = regulatory compliance
- −3 = worst case scenario

The indicators are equally weighted, so an average can be taken to obtain the result for the headline indicator, represented by *sectors* in the diagram (Fig. 1, below). Each sector is split into seven loci, each corresponding

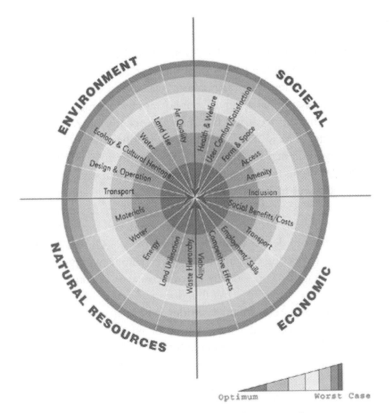

Fig. 1 Graphical chart of factors for assessing sustainability of any system or social organisation from community to nation.

to a level of performance: the outer loci (coloured dark grey) indicates the worst case performance (i.e. an average score of -3); the middle loci, known as the *median* (coloured light grey), indicates good performance, or regulatory compliance (an average score of 0). If the dark grey, inner loci is reached, the system has reached its *best case scenario* (an average score of $+3$).

LONDON

"London is today experiencing phenomenal growth ... This rapid expansion, of population and jobs, stems from London's exceptional dynamism, attractiveness and advantages in the new era of economic globalisation. It poses unique opportunities — but also challenges — if the potential benefits are to be maximised and the city's environment, quality of life and historic character are to be preserved and improved".

Ken Livingstone, 2002

London has received some severe criticism in its time — Sir Arthur Conan Doyle in his book 'Study in Scarlet' (1887) described London as 'that great cesspool into which all the loungers of the Empire are irresistibly drained'. 21st century Londoners complain about water and air quality, transport infrastructure, misappropriation of public money, immigration laws, low quality of life and the weather. But they are not the only city dwellers to have complaints — New York's operating deficit has risen to US$4.5 billion (based on governmental fund), with a total debt in 2002 of US$36.5 billion (McCall 2002); Mexico City has a 50% unemployment rate, some 8 million residents lack sewer facilities, and its indigenous people suffer poverty and racism (Miller 1998); Tokyo's narrow, congested roads have resulted in people being literally pushed onto trains, and the work force housed increasingly far from dense, centralised work places; Ciudad Juarez (northern Mexico) is expecting to run out of water within twenty years (Barlow 1999). The difficulties London has to overcome to become a sustainable world city pale in comparison, but this does not make them any less pressing; it is particularly relevant to consider this goal in terms of the city's growth patterns (in particular the creation of whole new cities within the London area), liveability and economic future, and its potential to continue to influence the direction of world cities in decades to come.

A SPeAR® assessment was undertaken using publicly available data to approximate how London is actually performing in terms of sustainability. It is important to note that this preliminary assessment, simply aimed at giving an idea of how London performs as a whole — it has not been rigorously undertaken.

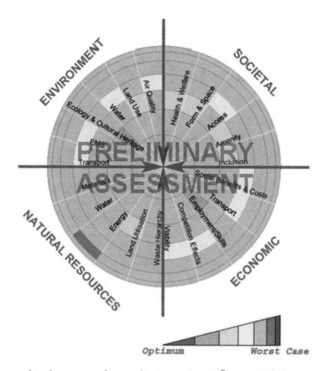

Fig. 2 The graphical output of a preliminary SpeAR® sustainability appraisal of London. The graphic format enables an understanding of the complex issues involved.

Immediately obvious is that the majority of the indicators are at the level of 'good practice', or better. Further, the quadrants are displaying comparable levels of performance, with the exception of the Natural Resources quadrant.

Of the four, the Societal quadrant is showing the best performance. This is an indication that some of the 'civilised institutions' spoke of early on are working well — notably London's schools, universities, hospitals and in particular, despite almost daily media criticism, the health service (reflected in the 'Health & Welfare' sector), and its museums and theatres are envied around the world. The people enjoying these facilities are as, if not more, diverse as in any city in the world — 50% of school leavers come from ethnic cultures, collectively speaking over 300 languages. Other sectors in this quadrant reflect the standard of accessible green space, the amenity and leisure facilities, level of community consultation and a scale of built environment compatible with people.

The Economics quadrant is performing well. Employment and Skills ranks so well as London has one of the most diverse and skilled work forces in the world, due to a focus on education within society as a whole. A strong and capable work force will enable the city to continue to generate wealth, as reflected in the Viability sector. The diversity of people coupled with the economic viability has attracted an assortment of businesses, shown in the Competition Effects sector (whether the ethics of London's enterprise culture is more 'piratical' than 'public service' is discussed in Chapter 17).

The quadrant performing the worst is that concerned with Natural Resources. This poor performance is partly because no base data was available to the assessors at the time of appraisal — improvement is impossible if the current state is unknown. However the City Limits report (www.citylimitslondon.com) which quantifies the ecological footprint of London has identified the nature of the problem in more detail, and has made available data on energy and resource use within London. It confirms that London's use of natural resources is poor — it has an ecological footprint the size of Spain (Chapter 5). It is worth noting that measures to tackle the problems have been proposed in the Mayor's strategies for waste, biodiversity, energy, air quality and ambient noise; the next few years could see a significant improvement in this quadrant.

Finally it might be surprising to note how well the Environment sector is performing. The highest rating sector here is Ecology and Cultural Heritage, which should not be surprising due to the city's long history and the proliferation of green space (Chapters 9, 11). Another perceived problem within the city is the quality of its air; nevertheless as a result of action being taken — including improved planning policies, in particular air quality management areas being established by most of the Boroughs — the performance of this indicator is in the regulatory compliance/good practice region. While current projections indicate air quality will worsen by 2010 as a result of increased vehicle use, they are likely to be overestimates — improved car engine design, along with practices and policies such as congestion charging may in fact allow London to reach Government and European targets by that time.

It is because of Londoners' intensive use of public transport, despite perennial complaints, that the air pollution is just tolerable and energy efficiency much better than many other cities. Every day 2.6 million tube journeys, 4 million bus journeys and 5 million train journeys bring 85% of Central London's employees to their workplace — moving this many people and still minimising environmental effects such as air quality and noise, and ensuring security and safety of the system is quite an achievement.

COMPARISON

Assessment tools such as SPeAR® indicate where a system is at a point in time, and can identify targets for improvement. A benefit of continually using the model over time is that it can be used to monitor the system's performance, and identify successes in meeting different targets. It is also possible to use this approach to compare systems, and in this example cities. A third-world city particularly relevant at the moment is Johannesburg see Fig. 3. Again, this is not a rigorous assessment.

Unsurprisingly, Johannesburg is not performing as well as London in any of the quadrants as a whole. However in some individual sectors it is performing as well, and in the Water sector of the Natural Resources quadrant it is actually performing better. This is because Africans are fully aware that water is a limited and valuable resource, therefore water quality standards

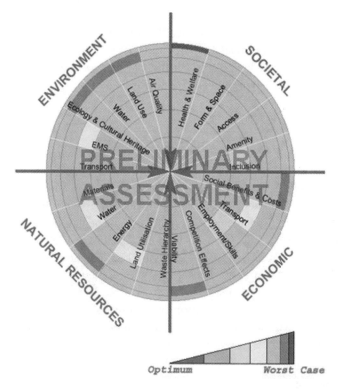

Fig. 3 The graphical output of a preliminary SPeAR® sustainability appraisal of Johannesburg. The graphic format enables understanding of the complex issues involved.

have been more stringent than in Europe, and better enforced than in most places in the world.

Two sectors are prominent in their good performance. Tribal traditions of Africa result in the involvement of the whole community in any decision making process, reflected in the Inclusion sector. The Employment and Skills sector reflect the aspirations of the new South Africa, which have been laid down in its Constitution, including provision of training and access to employment.

Some frightening issues in Africa are reflected in the Health and Welfare sector. The spread of HIV/AIDS has had horrifying consequences that affect all aspects of life and work. The prosperity of Johannesburg and all of Africa is seriously threatened. The susceptibility of HIV/AIDS sufferers to Air Quality problems has been exacerbated by the use of coal to heat and cook. Aside from the energy efficiency problems, the air-borne coal particles have given rise to a high percentage of the population having upper respiratory diseases. A workforce that has been reduced in number and efficiency, mainly due to ill-health, further intensifies the strain on health care provision (Social Benefits and Costs sector). Combined, these factors make it almost impossible for the city to evolve a strong economy in the current circumstances (Viability sector).

CONCLUSION

It has been argued here that useful conceptual frameworks can be devised for assessing sustainability issues for activities and communities within cities, and for cities as a whole, that will identify priorities for action and show the benefits of dealing simultaneously and in a connected way with different aspects of the environment, society and economics e.g. air quality/health/viability.

This modelling could be used to build a database not only for comparison of cities, but as a way of identifying best practice from around the world, information that is not available at the moment.

London is indeed a first class city of which Londoners should be proud! As we continue to build upon our strengths and address the weaknesses, London will not only remain one of the best cities in the world, it will no doubt continue to be an example to others.

REFERENCES

Anon (1964) *The Concise Oxford Dictionary*, Oxford University Press.
Felipe Fernández-Armesto (2000) *Civilizations*, Macmillan Publishers Ltd.

World Commission on Environment and Development (1987) *Our Common Future*, Oxford University Press.

HMG (1999) *A Better Quality of life: A Strategy for Sustainable Development for the UK*, The Stationery Office.

HMG (1999) *Quality of Life Counts: Indicators for Sustainable Development for the United Kingdom: A Baseline Assessment*, Government Statistical Service Publication.

DEFRA (2001, 2002, 2003) *Achieving a Better Quality of Life*, Review of progress towards sustainable development — Government annual report.

World Commission on Environment and Development (1987) *Our Common Future*, Oxford University Press.

Herman, E. D. (2002) *Sustainable Development: Definitions, Principles, Policies.*

International Institute of Sustainable Development (1996) *Who developed the principles?* http://www.iisd.org/measure/principles/2.htm.

Sir Arthur Conan Doyle (1887) *A Study in Scarlet* (published by Oxford University Press in 1993).

McCall, H. C. (2002) *2002 Comptroller's Report on the Financial Condition of New York State.*

Miller, G. T. (1998) *Sustaining the Earth*, Wadworth Publishing Company.

Barlow, M. (1999) *Blue Gold: The Global Water Crisis and the Commodification of the World's Water Supply*, International Forum on Globalization.

Chapter 17

A London Fit for Pirates

Jonathan Glancey

Born in Indiana in 1871, Theodore Dreiser started his career on the Chicago Globe and St. Louis Globe-Democrat in the early 1890s. He watched and reported on the shady financial dealings and grubby politics of these boisterous cities.

These became the stuff of his future novels, and specifically of the trilogy that began with the publication of his third novel, 'The Financier' in 1912.

Set in Philadelphia — the city of brotherly love — of the mid 19th century, it depicts the chaotic and changing circumstances of the free-booting American financial system of that period, whose principal character, Henry Worthington Cowperwood, was, to quote Dreiser, 'a financier by instinct, and all the knowledge that pertained to that great art was as natural to him as the emotions and subtleties of life are to a poet.' The life of the fictional Cowperwood was based on that of the all too real Charles Tyson Yerkes whose high-minded Quaker ancestors had emigrated from Wales to America in 1682. Charles Yerkes was a very modern hero in the Thatcher–Blair mould: a broker and underwriter in his twenties specialising in the sale of government, state, and city bonds. He was the master of putting together consortia to make money from nothing, yet along the way he helped shape tramways and subway systems in Chicago, New York and London. He was jailed for the embezzlement of public funds in Philadelphia in 1871, and was later to flee the United States.

His business methods did not endear him to everyone. They included bribery of council officials and, when that didn't work, the use of 'professional vamps' to seduce and then blackmail Chicago's law-makers. At one stormy meeting, people turned up with nooses and guns hoping to string him up like a common pirate. Stock he sold to investors for $100 million was worth less than $15 million four years later and he admitted to friends that his usual way of working was 'to buy up old junk, fix it up a little and unload it upon other fellows'.

I like the story of the way he got the final part of the Chicago elevated railroad — the 'L' — completed under threat of tough penalty clauses. As the deadline drew near just after Christmas 1899, the city's public works commissioner declared the line not just incomplete, but unsafe: to cut corners and to increase profits, portions of the structure lacked the specified number of rivets. Operations were suspended immediately. Yerkes had other ideas. He ordered the trains to operate anyway and the next day at Wrightwood station, four policemen arrested the driver of the first train of the day. One of Yerkes's directors, Frank Hedley, took the controls and continued, illegally, towards the Loop only to find 50 policemen lined up across the tracks, blocking his path. Instead of stopping, Hedley sped his pirate train towards Chicago's finest, and scattered them. It was like a scene from the Keystone Cops. Lawmen further along the line laid timbers across the tracks, and Hedley was forced to stop. Was he arrested? Not a chance: too many Chicago businessmen had fingers in Yerkes's pie. He was granted an extra five months to bring the works up to scratch and finish the job.

This took the handing out of over a million dollars in bribes to local businessmen, lawyers and politicians. But, this was, even by the crooked standards of the day, no way to run a railroad. Before the law laid its heavy and corrupt hand on his shoulders, Yerkes made tracks for New York City.

Here, he sold the bulk of his transit holdings, before sailing for England. In London, he invested in the development and electrification of the new Underground railways. Applying much the same methods he had adopted in Chicago. Yerkes got some of the work done, but when he died in 1905, the embryonic Underground was on the verge of bankruptcy.

Not that he had suffered. Like one of today's less colourful British fat cats, Yerkes, whatever the distress he had caused to other people, to their public bank-accounts, institutions and cities, did well for himself. Who cared if the businesses failed or even if the trains came off the lines or crashed? Money was the goal. In New York, Yerkes had furnished his

$1.5 million mansion on Fifth Avenue with a marble staircase, a conservatory stocked with free-flying birds and a gallery full of European art treasures. He built a second mansion, a few blocks away, for his favourite mistress.

How did he do it? Or, more to the point, how did he get away with it? By using methods very similar to today's Public Private Partnership (PPP) and Private Finance Initiatives (PFI) policies, golden instruments for today's robber barons and commercial pirates, who working hand-in-greasy-hand with a socially amoral Treasury, truly believe that the spirit and questionable methods of the buccaneer can be placed in public service. Nearly a century after Yerkes's death, the British government is introducing PPP's to revitalise the London Underground, which has an American boss who is opposed to the whole idea. The career of Charles Yerkes is, to say the least, topical.

A pirate who sailed the high seas of high finance, and for whom booty was the ultimate port of call, Yerkes would be more than welcome in London today. And, yet, we can hardly blame Yerkes for what he believed. After all, businessmen in Britain and America were often encouraged to become pirates — privateers — in the service of the state. Francis Drake, one of the most successful of all pirates, remains a national hero. He was knighted for his efforts by a grateful Queen Elizabeth in 1580. Between 1660 — when the Stuarts returned to the throne — until the bursting of the South Sea Bubble in 1720, and slightly beyond, piracy was positively encouraged. It was a language British businessmen understood. They were backed by kings and governments who believed in more or less untrammelled private enterprise.

Henry Morgan — a famous pirate of Captain Kidd's and Blackbeard's day — retired with a knighthood, as deputy-governor of Jamaica. The system, though, could bite back; and just as Yerkes was jailed for embezzlement in Philadelphia when he overstepped the mark, Captain Kidd, who steered all too close to the wind, was hung in chains at Execution Dock, Wapping. Significantly, Kidd had begun his career as a New York merchant, and sea captain. He had been commissioned by King William III who referred to him as our 'beloved friend, William Kidd.'

Today's pirates are more fortunate than Captain Kidd. There is no death penalty, while booty pours into their pockets without them having to step aboard a creaking ship. Senior bosses in railways and public utilities commonly receive vast bonuses, sometimes running into millions, to encourage them to stay aboard, even if their record is poor.

These kinds of payments — bonuses and so-called golden parachutes — have been paid even when the culprits are on their way out, and the company's performance has been recognised as disastrous by passengers, the media and shareholders. One transport executive received over £1 million in severance money in such circumstances. Another received a pay rise of well over 50 per cent even though lives had been lost on his watch thanks to company negligence. After a fatal crash, other directors pocketed huge bonuses, too.

John Edmonds, leader of the GMB general union, has condemned 'the staggering greed' of what he referred to as 'PFI privatisation pirates', adding that 'public sector workers cannot be expected to exercise restraint when they see millions of pounds of public money squandered on salary hikes for directors rather than being used to revitalise our schools and hospitals.' It is fascinating to witness the trade union movement regain the moral high ground; it doesn't seem so long ago that unions were accused of what to all intents and purposes was piracy: holding the nation to ransom with their fearsome and unreasonable demands. The pirates are now the business community encouraged as they have been since Drake and Good Queen Bess by the state. And, yet, government and business today have the bare-faced cheek to say that their Yerkes's-style methods of investing in the infrastructure of our cities are for the public good. At least Blackbeard never claimed as much, and went down fighting, cutlass and pistols in hand. 'Come', said he to his skull-and-crossbone crew: 'let us make a hell of our own, and try how long we can bear it.' Now we plan — if that's the word — to make a purgatory of our cities — while pretending we're doing it for the right reasons: the public good.

Before he was hung at Tyburn in 1725, Jonathan Wild, the infamous thief-catcher who traded in received goods, who advised the government on crime and caused sixty-seven of his fellow villains to die on the scaffold, penned a public letter from prison advising his successors in how to succeed in dirty business where he had failed. 'The public good', he wrote, 'which has ever been the mask of self-interest and private avarice, must always be on the tip of your tongue. This notable phrase is swallowed down by the multitude with great approbation, and they turn their eyes with reverence upon the man who only makes the mean external show of it. They cannot be made to think ill of the person whose favourite topic is the welfare of his country, not withstanding his more secret intentions are upon the most selfish principles in nature; for who can imagine that he who constantly brawls forth his good sentiments, his esteem and affection for his fellow creatures, and is at every juncture wracking his brains for schemes and plans for their benefit should have any such principle as his own interest at heart.'

Exactly, so. How we are led by government and our own swinish folly and fears to suspend our judgement and to believe in those piratical evils of our age, PFI and PPP. We have boarded the public realm over the past twenty years, and robbed, raped, cutlassed and plundered it with hypocritical savagery.

We no longer talk much about the design or infrastructure of public services, the quality of their engineering and architecture; we read only of failures and the exploitation of labour, of train drivers, once heroes to generations of schoolchildren being treated like powder monkeys, of firemen being paid the equivalent of a cabin-boy's wages, and being told that they ought to get on their bikes and earn a bit of extra cash by working on their rest days. What we like talking about is money, profit, golden handshakes, pension funds, house prices, equity, shares: booty, bounty, private treasure troves. Gold, doubloons, pieces of eight.

We have returned to being a nation of pirates — after what was a brief and uncharacteristic spell of common sense, decency and public spiritedness, in London, say, from the creation of the London County Council at the end of the 1880s to the rise of Thatcherism from the ashes of faded, jaded, wrong-turning socialism. Having plundered and trashed London Transport, for example, once the finest, integrated public transport system in the world, we expect today's Captain Kidd's, Jonathan Wild's and Charles Tyson Yerkes to sort it out through cynical, cheap and amoral private finance initiatives.

Significantly, though, London Transport was first pulled together and reached its peak under the aegis of Lord Ashfield, a savvy businessman brought up in the New York of Yerkes' time, and appointed chairman of the new London Passenger Transport Board (LPTB) — a public corporation like the BBC in 1933 — and the remarkable Frank Pick, his high-minded, incorruptible chief executive. Pick made London Transport work; he made it profitable, he made it beautiful, an instrument of efficiency and democracy; he did not earn bonuses; he did not live to pensionable age; he was nothing like a fat cat, much less a pirate. He would not take a knighthood much less a peerage. The architectural historian Sir Nikolaus Pevsner described him as Lorenzo the Magnificent of the 20th century. Sir Kenneth Clark — Lord Clark of Civilisation — said 'In a different age, he might have become a sort of Thomas Aquinas.'

What this? London Transport, Renaissance Florence and medieval scholastic philosophy in one and the same breath? It seems hard to believe now (Glancey 2001). Shortly after his death in 1941, a last essay of Pick's was published in the Congregational Quarterly. He was recalling a holiday walking through the Swiss Alps; 'all at once, at about 10 000 ft above sea level, there seemed to come an extraordinary shift of colour. The blue band of the sky had come down to earth and was enveloping the land. It seemed to me as though Heaven had come down to earth. Above, the sky had almost a hard whiteness. The whiteness of God. We were coming face to face.' I can't imagine modern PPP/PFI pirates sharing this experience, nor expressing themselves quite so lyrically.

Here, by way of contrast, is Charles Tyson Yerkes again, just before he died: 'I liked to buy up old junk, fix it up a little, and unload it upon other fellows.' Like the London Underground.

In seafaring terms, where Pick was a fair and strategically-minded captain in a ship of the line fighting with Nelson at Trafalgar, Yerkes, a businessman for our piratical times, was one with Blackbeard and Captain Kidd. I can't help thinking that our natural English — or is it British? — sympathies are with Blackbeard and Yerkes rather than God and Frank Pick. And, even then, which of us hiding behind a rosewood veneered desk and a £750 suit has the sheer courage of the pirates of yore? In the words of W S Gilbert:

> *'Oh better far to live and die*
> *Under the brave black flag I fly,*
> *Than play a sanctimonious part*
> *With a pirate head and a pirate heart'*

Pirates, though, whether in Penzance or in the boardrooms of the City of the London, will be … pirates.

So, what is the point of even beginning to discuss issues of urban ecology, of how we might envision the City Resplendent, just cities for people of all backgrounds to flourish and prosper in? Surely, these can only be the stuff of fiction, given that we have chosen to raise the Skull and Crossbones and set sail along the corrupt channels of PFI and PPP out into the short-term, fast-buck, free-booting sea of high finance and government-approved piracy?

To end, let me sail back to port: in The Stoic, published in 1945, the last in Theodore Dreiser's trilogy, he describes the ultimate attainment of his hero — our hero today — based on Charles Tyson Yerkes: 'At the height of his success, when he had settled old scores and could easily have become the smiling public man, he chose instead to rip the whole fabric of American civilisation straight down the middle, from its economy to its morality. It was the country that had to give ground.'

Dreiser died a few months later; a lifetime of observing such greed and folly had made him increasingly left-wing as he grew older. In the late 1930s he had gone to Spain to support the socialist government against Franco. On his deathbed he joined the Communist Party, and who, at the time, could blame him?

A half century on, we are unlikely to be tempted to set our course quite so extreme left, but until we learn to rid ourselves of piracy, to stop being so very selfish, short-sighted and greedy, our environment and our future is in hands as unsafe as those of Blackbeard, Jonathan Wild and Captain Kidd.

REFERENCE

Glancey J. (2001) *London Bread and Circuses*, Verso, London.

Chapter 18

London's Governance and Sustainability*

Tony Travers

Over the last 250 years London has suffered from confused government and a poor environment and at the same time its population has increased by a factor of 11. In the late 18th and early 19th centuries the expansion of London was uncontrolled, causing a lack of infrastructure and municipal services in many districts of the city. (Further aspects of London's history are discussed in Chapter 4 and by Aykroyd (2001).) The city was governed comparatively well by the City Corporation but the rest of London was controlled by parish vestries, which were increasingly ineffective. London's environment and its impact on public health were the main driving force for the reform of this situation. London's water was drawn by private companies, unfiltered from the same stretches of the Thames into which soiled water was discharged. This led to outbreaks of typhus and several outbreaks of cholera, killing thousands of people at a time. The main campaigner for reform was Edwin Chadwick (1800–1890), who proposed a central commission to manage the water, drainage, paving and street cleaning for all London. However Parliament, in 1855, chose a more decentralised approach, in which a lower tier of elected vestries and district boards elected a central Metropolitan Board of Works (MBW). The MBW had very limited powers but achieved its main purpose — the building of a main drainage system. Following the 'Great Stink' of 1858, when sittings

---

*Background material was also contributed by Sam Hunt.

at the House of Commons had to be abandoned because of the smell from the Thames, the MBW was allowed to get on with this work independently, and by 1865 the system was in place.

Less progress was made in the improvement of air quality. In 1853 the Home Secretary, Lord Palmerston, pushed through a Smoke Nuisance Abatement measure, which had little effect because it only applied to industrial chimneys, not domestic ones. Fogs, such as those painted by Monet (see Chapter 8), only began to become less frequent in the 1890s because, it is thought, the expansion of the suburbs spread the domestic sources of smoke over a wider area.

Following revelations of corruption, the MBW was replaced in 1889 by the more powerful and accountable London County Council (LCC) and the parish system was replaced by a lower-tier of metropolitan boroughs. The LCC delivered significant improvements in transport and housing in the late 19th and early 20th centuries. However, outside the LCC boundaries, local government was an incomprehensible jumble of counties, districts and county boroughs where the London conurbation continued to expand.

In 1965 much — though not all — of London's built up area was embraced by the Greater London Council (GLC) and a system of 32 lower-tier London boroughs was created within its boundaries. The GLC lacked effective power. From its earliest years there were many who sought to weaken or destroy it, especially those who disapproved on political or financial grounds of the popular: municipal initiatives and campaigning of its leader Ken Livingstone. In 1986 the GLC was abolished by Mrs Thatcher's Conservative administration. Services run by the GLC were re-distributed to Whitehall, government-appointed boards, city-wide committees of boroughs and to the individual London boroughs.

The years from 1986 to the re-creation of London-wide government in 2000 saw the evolution of an extraordinary civic coalition, involving the boroughs, private companies and voluntary organisations. Some of this cooperation was encouraged by central government after 1992. A new body was set up to lobby for the capital — London First — and, in 1994, the Government Office for London was created. Many of the capital's new civic class believed that in the longer-term (i.e. when, and if, the Conservatives lost office), a new elected government for London should be created.

In its 1997 manifesto, Labour proposed that London should have a directly elected mayor, overseen by a small assembly. Ken Livingstone was elected in May 2000 as the first-ever elected mayor of London and since then the Greater London Authority (GLA) has become a key element in London's political establishment.

The Greater London Authority Act of 1999 gives the Mayor the responsibility of regularly assessing and reporting on the state of London's environment. Ken Livingstone has stated that he has a vision of London as a sustainable city and aims to make significant improvements in London's local environment as well as reducing its impact on the global environment. To this end, air quality and biodiversity strategies, as well as draft energy, waste and ambient noise strategies have been published (see Chapter 5). However it is uncertain whether the Mayor can achieve the objectives of these strategies, given the present extent of his powers.

The main functional bodies through which the GLA can implement its strategies are the London Development Agency (LDA) and Transport for London (TfL). The LDA has a business-led board, appointed by the Mayor, responsible for preparing the Mayor's business plan for the capital and has a budget of over £300 million a year. It aims to develop London's economy, whilst encouraging sustainable practices in the business community. TfL aims to improve the capital's air quality and energy use by measures such as replacing its bus fleet with cleaner vehicles. The Mayor also aims to improve air quality and energy use through the congestion charge for central London. The Mayor has been given powers to implement his municipal waste strategy, ideally in partnership with London's waste authorities, but if they do not cooperate he can impose his will by various means. The GLA's role in implementing its biodiversity and ambient noise strategies is largely one of campaigning.

Local roads, street cleaning and lighting, local planning and refuse collection and disposal are all the boroughs' responsibility. The Mayor and assembly are just part of the crowded and complex world of London governance. For instance, improvement of the street environment in central London is the responsibility of the boroughs, the Royal Parks Agency and the Central London Partnership as well as the GLA. The Mayor of London cannot simply impose his will in the way say, the mayor of New York or Paris might.

As well as lacking power, the Mayor lacks financial resources. The GLA and the boroughs together only collect £3 billion of the £60 billion paid by Londoners in tax each year and about a third of this local tax goes to the GLA. Overall public spending on London is in the range of £45–50 billion, meaning that Londoners contribute £10–15 billion to the rest of the country. Earnings per head in the capital are well above the national average but London is a city of intense poverty, whose basic assets, such as railways, water and sewage systems need to be renewed. Also, even though London's population has risen sharply during the past 10 years, its spending share has declined, as shown by the table below (based on government figures).

Year	1995–96	1996–97	1997–98	1998–99	1999–00	2000–01
London's share of public spending in England (%)	18.0	17.5	17.4	17.1	17.2	16.5

Many aspects of London's local environment are worse than those of smaller urban areas outside London, and its waste discharges to other areas are increasingly unsustainable (c.f. Chapter 5). But when its large population is taken into account London is not proportionally a major contributor to global pollution and in many ways it contributes to sustainability. Firstly London is an efficient user of land, with a population density of 4600 people per square km compared to the UK average of 245 per square km.

Secondly, car use in central London is considerably lower than the UK average, though in the suburbs it is worse. The average annual distance travelled by car per person is 3376 miles in London and 5524 miles in Britain (DETR 1999). 60% of people travel to work by car in London (46% in inner London) compared with 71% in Britain (DETR 1999). This implies that the capital's transport has a smaller impact on fossil fuel use and carbon dioxide (CO_2) generation than that of the rest of the country. However low overall vehicle speeds increase London's relative (and absolute) contribution to nitrogen oxides (NO_x) and fine particle emissions. Energy use in London was 21 megawatthours (MWhrs) per person in 1997 whereas the national average was 29 MWhrs per person (GLA 2003). This reflects the lack of energy intensive industry in the capital, as well as economies of scale resulting from a higher population density, with many people housed in flats. Average household energy consumption per person is higher in London because of its above average affluence and smaller average household size.

London produces 27 kg of municipal waste per household per week, which is slightly above the national average of 26 kg (GLA 2002). However the capital's role as a shopping, tourist and business centre means that a higher fraction of this waste is from non-household sources. 11% of municipal waste in England is non-household but in London it is much larger, being about 24%.

These statistics help explain London's and Londoners' impact on the UK's environment. The perception of London as a major source of pollution is largely based on measuring its absolute impacts. It seems that London's 'ecological footprint' would be even larger if its 8 million population behaved like other UK communities, which are of course spread out at much lower densities. Nevertheless the capital's impact on Europe's environment could be greatly improved.

The growth of London and other large world cities are an inevitable feature of global development, as the World Bank and many other bodies have noted. The main issues for their inhabitants and visitors are how to ensure that they function efficiently and that their populations have a better quality of life. The only answer is good government. Politics must be used to trade-off competing demands and the authority and 'pulpit' role of government must be used. However, in London the Mayor and GLA have limited influence and city wide approaches are very difficult with London's government so weak and fragmented.

However progress has been made by collaboration between all levels of Government, including the national Government, as was evident with the successful introduction of the congestion charge. This gives some hope that problems will be tackled in the end, even if quite slowly!

REFERENCES

Aykroyd, P. (2001) *London — A Biography*, Viking, London.

Department of Environment, Transport and the Regions (1999) *Transport Statistics for London 1999*, pp. 18–21.

Greater London Authority (2003) *The Mayor's Draft Energy Strategy*, Chapter 2, p. 12.

Greater London Authority (2002) *The Mayor's Draft Municipal Waste Management Strategy (Public Consultation Draft)*, Chapter 2, p. 11.

Chapter 19

Mayor of London's Vision of the Future of London's Environment

Ken Livingstone

INTRODUCTION

This book, based on the LEAF conference in September 2002, addresses a host of problems, putting some new issues on the agenda, bringing new slants to some familiar problems, and I'm glad to see real solutions being offered to some of the critical issues. I am impressed by the wide range of issues that were covered from environmental health, water resources, values of green space, risk of flooding and problems of climate change. And I am particularly glad to see that the conference considered many of the detailed problems which I have addressed in my environmental strategies for London. I would like to congratulate all those responsible for making the conference happen. It was a valuable opportunity for everyone concerned with London's environment to take stock and think about where we are going.

The book provides an extremely valuable analysis of where we are at present — a snapshot of London's environmental problems. But where do we go from here?

WHAT ARE THE CRITICAL ISSUES?

Londoner's no longer suffer the pea soup smogs of the 1950s which I remember so well, but London's atmosphere is just as insidious today with the high levels of nitrogen oxides and particulates which make London one of the most polluted cities in Europe. We have to crack this as a matter of urgency. My air quality strategy launched in September 2002 makes a large number of proposals to reduce the emissions from the most heavily polluting vehicles, for it is mostly road traffic that is responsible for the problems.

A second major issue is how we deal with waste. My Municipal Waste Management Strategy, published in 2003, calculates that as a result of our everyday activities we produce 17 million tonnes of waste every year, *most of which is literally wasted.* At present we send most (70%) of our household rubbish to landfill way outside London. We cannot continue to do this and we have to find ways of converting the vast majority of the wastestream into products. We have already been moderately successful through our grant of £5 million to set up London Remade, which is dedicated to doing just this. We have a long way to go. At present London recycles about 9% of its household waste and I want to see this increased as rapidly as possible. Again our performance is just about the worst of any major European city. Because of the lack of concerted action across London after the Greater London Council (GLC) was abolished in 1986, we are starting from a long way behind others in the field. But I am determined that London now gets its act together using new technologies to crack the problem. My waste management strategy makes detailed proposals for how we can get things moving. I am delighted that we have been able to provide funding for a host of new initiatives across London through allocation of £50 million from the London Recycling Fund. This is now making a real difference and will significantly improve our recycling capacity.

Energy is at the root of many environmental problems and I am delighted to have published London's first regional energy strategy, which may well provide a model for other regions of the UK. Its main aims are to address social issues of fuel poverty and to find ways in which London can reduce its reliance on fossil fuel energy. In all this I am acutely aware of the impact London has on global climate change. We need to find solutions which meet these wider global problems as well as looking for cleaner forms of fuels which will reduce our air pollution problems here at home. We have made a great deal of progress over the past two years establishing an overall Energy Partnership for London; also bringing together all the key players to set up a Hydrogen Partnership to promote and develop new technologies such as fuel cells. I was pleased to launch a pilot scheme of three fuel-cell powered buses in January 2004.

WHERE DO WE GO FROM HERE?

Many of these issues are addressed in other chapters. My job is to ensure that we build the best available knowledge and experience into our decision making, setting a strategic framework for dealing with London's environmental problems over the next 20 years.

But it isn't just a matter of ensuring that we improve London's environment — although we all want that. What I am trying to do is raise these issues across a wide agenda including economic development, social inclusion, health, and sustainability. My vision is to ensure that London is an exemplary sustainable world city. That vision lies at the heart of my London Plan published in 2004. I am delighted that my proposals for sustainability received widespread support during public consultation on this plan. What this means in practice is developing new skills and new technologies to ensure long-term environmental sustainability and at the same time making a real contribution to economic progress and especially making London a city that people want to live in.

I said at the time that the conference was very timely coming as it did close on the heels of the Johannesburg Summit (WSSD). Earlier there were some very positive outcomes from the Earth Summit in 1992. One of these was the very clear demonstration at the Local Government session that it is major cities that provide the key to long-term sustainable development. Many cities are already achieving far more than their own central government's targets. There was also a great deal of optimism about the future amongst Local Government representatives from all over the world, who showed numerous examples of how solutions are actually being delivered on the ground. We in London are playing our part in this in addressing our immediate environmental problems. This involves the development of a host of new technologies and new solutions.

I announced at the conference that one way I would like to take things forward from Johannesburg, here in London, is to build on the Green Procurement Code that we have already developed. Considerable progress has now been made to explore with the public sector and key elements of business how we can develop a London-wide programme for green procurement, using the buying power of all our public sector bodies, which is immense.

Another message from Johannesburg was that the rapidly growing cities of the developing world are faced with enormous environmental problems and need to find ways in which they can develop their sustainability. Over half the world's population now lives in towns and cities and by 2020 two thirds of the world will live in cities, with much of the rest depending on urban markets for their economic survival. I have argued previously that there will be no sustainable world without sustainable cities.

Finding ways of making modern cities vibrant and liveable, with sustainable economic growth, whilst at the same time making them self-regulating and reducing their global environmental impact is one of our greatest challenges. Most western cities are inherently unsustainable. They are dependent on vast quantities of resources which we consume every day. Although cities take up only 2% of the worlds land surface they use over 75% of its resources.

The report co-funded by the GLA on London's Ecological Footprint published in 2002 demonstrated that London is dependent on an area 293 times its actual size. This footprint is of course dispersed around the world and has ecological impacts in many different ways that are difficult to assess. What it does demonstrate very clearly is that instead of seeing cities as the source of the world's environmental problems, we can look at them as the most efficient way of contributing to the solutions. To put it simply, it is in cities that the greatest opportunities exist to make the necessary changes towards sustainability. There are enormous potential savings to be made through the economies of scale inherent in higher urban densities, and this lies at the heart of my Spatial Development Strategy — the London Plan.

The Local Government Declaration at Johannesburg recognised that we all have a part to play in finding practical solutions which can be applied globally. My vision of London as a sustainable world city means that, as well as putting our own house in order, it is possible that through our investment opportunities we are able to develop workable solutions which could have much wider application in the developing world.

I referred earlier to the successful London Remade project with £5.4 million input from the London Development Agency (LDA), which aims to create new markets for waste, encourage private sector investment, and create local employment. It will make more efficient use of natural resources and reduce the amount of waste that ends up in landfill. Considerable interest is already being expressed in this kind of project.

This shows how all my strategies and policies need to address sustainability. I announced at the conference in 2002 that I had established a Sustainable Development Commission for London. I am glad to say that this Commission has made enormous progress over the past two years identifying ways in which we can put sustainability into practice. The Commission published a Framework for sustainable Development in London in 2003 which provides a template for decision makers to ensure that these issues are addressed across all spheres of policy development in both public and private sectors. Although the GLA has a duty to promote sustainable development, we will need to work closely with our partners and stakeholders if we are to achieve real changes. That means many actively engaged in the environment of London, including those who came to the LEAF conference. I see the new Commission having a vital role in forging a real partnership for London's environment.

Another important initiative since the conference was the publication of my State of the Environment Report for London. This report provides a snapshot of the current position regarding environmental quality and includes measures of environmental performance. It will be invaluable as a baseline to monitor our success in implementing new environmental policies and programmes.

CONCLUSION

I would like to conclude by emphasising that as individuals we need to be aware of environmental consequences in every decision we make, in the way we travel to work, or the way we deal with our household rubbish. Influencing lifestyle changes and helping to forge new value systems will be fundamental in helping London to lead the way to achieving real improvements in the environment and hence quality of life for Londoners. I believe that this book, based on the LEAF conference, is important in providing a clear focus for practical action for the future. We all share responsibility and it is up to everyone to play their part.

Postscript

The political noise on climate change and sustainability is now deafening. In the UK at least there is a consensus about the need for more urgent action. There is a growing realisation that, with increasing urbanisation, the role and function of cities and regions is pivotal. As the Mayor wrote in the foreword, sustainable development won't succeed without sustainable cities.

London's challenge is to demonstrate what this means on the ground, as many of the authors in this volume have done, and to show leadership over the coming years to accelerate the effort to tackle climate change.

The immediate political context in the UK is favourable. In Autumn 2004, high profile speeches by Prime Minister Tony Blair, his Chief Scientific Advisor, Sir David King and the Leader of the Opposition pushed climate change further up the domestic agenda.

The Prime Minister indicated that climate change would be one of the Government's two main themes during 2005, when the UK will Chair the G8 and take on the Presidency of the EU.

In London, the Mayor is now building on his first term achievements. Following his election victory in June 2004, he announced plans to implement a Low Emission Zone to clean up emissions from lorries, coaches, buses and taxis. The primary aim is to tackle air pollution. This will cover the whole of greater London.

The congestion charge has brought about a 30 percent drop in congestion in central London and resulted in a massive modal shift to public transport, especially buses. Initial monitoring shows a 19 percent drop in CO_2 emissions from traffic inside the zone.

Seventy percent of London's CO_2 emissions, however, come from buildings.

To meet this challenge the Mayor is establishing a London Climate Change Agency. Its primary purpose will be to meet the ambitious greenhouse gas reduction targets outlined in the Mayor's Energy Strategy.

The Agency will complement the Mayor's strategic planning role and engage with the private sector to develop projects promoting a sustainable and secure energy infrastructure, especially in the opportunity areas such as the Thames Gateway and the London Olympics.

Locally distributed combined heat, power and cooling will be promoted wherever viable. London has 27 percent of the UK's CHP potential. Transitional use of natural gas will be replaced with a growing proportion of renewables and renewable hydrogen as they become more cost effective.

The Agency will also promote energy efficiency for London's existing as well as new corporate, institutional and domestic building stock. This will supplement work with developers to achieve new standards in sustainable design and construction. An initial focus will be on the GLA Group building portfolio.

The Mayor has appointed Allan Jones MBE, the former energy pioneer in Woking, just outside London, to run the Agency.

Woking has undertaken groundbreaking work on energy services, CHP, fuel cells and renewable energy, water efficiency, environmentally friendly waste recycling and energy recovery systems and alternative transport fuels. Its Borough Council received the Queens Award for Enterprise, the first time a local authority has received this award, for the development of local sustainable community energy systems.

London has a dynamic metabolism. We are a growing city. But, as David Goode discussed in his contribution, we can take action to reduce our ecological footprint. As Deputy Mayor, the Mayor has asked me to lead on environmental issues and accelerate our effort on climate change. The scale of this vision has to be matched by innovation and action on the ground. Cities have real potential for change. It is London's task to show how.

Nicky Gavron AM
Deputy Mayor of London

APPENDIX

Reports of Breakout Groups at LEAF Conference 2002

As part of the LEAF conference 2002, the delegates were invited to join one of ten separate 'breakout groups' to discuss different aspects of London's environment. The topics of the groups were (1) Sustainable development in London; (2) Sustainable buildings; (3) Re-engineering the household — How can we make recycling as easy as putting dirty washing in the laundry basket?; (4) Ecology of London's waterways; (5) Changes in avian biodiversity (including sparrows); (6) Climate change and London; (7) Monitoring the climate of London; (8) Monitoring and reducing noise; (9) Lagoon of London; and (10) Young people's perspectives on London. Each breakout group was led by a convenor. Their summaries of each group's findings are given here.

SUSTAINABLE DEVELOPMENT IN LONDON

John Murlis

This was a well-attended group of diverse participants. There were teachers and students, consultants and managers, government officials and members of public groups. Many sectors were represented, including waste, transport, conservation, education and information services.

The expectations of this group were that they would gain a better understanding of sustainable development, including ways of turning the intangible concept into something concrete, and the priorities for the future. Participants also wished to share practical experiences and to consider how best to provide citizens with the information they needed to act more sustainably.

We considered first what made London a particularly attractive place to be in. Participants described the excitement of living in a World City with its heritage, culture, vibrancy and energy, diversity and opportunities for employment. They valued the river and the open spaces of London, its gardens and architecture. The transport system was appreciated, when it worked, and people mentioned the pleasures of walking and cycling and the easy access to countryside. Finally the group participants spoke of the sense of place and community in London and the feeling of being at the centre of affairs.

However, participants also felt that there were many challenges to liveability in London. In particular, traffic, with congestion, noise and pollution, and the cost of travelling were felt to be generally unacceptable. Other disagreeable features are crime, litter and waste, pronounced social inequity and the general cost of living. The developing culture of London as a city that works "twenty four hours a day, seven days a week" contributes to excitement, but also places pressures on liveability. It was agreed that government in London does not seem to be joined-up and that there is a lack of long term planning. The alienation of citizens in London from government and social dislocation were considered to be an obstacle to good planning (see Chapter 15).

We then explored the connections between all these aspects of liveability, it and the ideas and practice of sustainable development. All elements of sustainable development were considered including, social, economic and environmental and how they needed to be combined to improve the overall quality of life. Many of the linkages between these elements are difficult to resolve, for example disadvantages of noise and pollution of traffic and the division of communities by roads arise from economic pressures for greater mobility. Furthermore there is an increasing demand for transport across and into London because of the lack of availability of affordable housing near work places.

This suggested to the group that the problems faced by Londoners would need solutions that dealt with environmental, social and economic aspects together. As we turned to solutions we considered first the institutional machinery through which they would be delivered, including different levels of government, non-governmental organisation, communities and families.

Participants believed that there is a generally low level of public engagement in local government and that there is much to be done to build a sense of belonging and ownership of measures that would deliver liveability. Although there is scope for policies and projects to encourage a more sustainable pattern of living, it was felt that a 'bottom-up' approach, depending largely on education and learning might be more effective in the long run. This process would engage the education institutions, parents and local groups as the main learning communities. A particular need identified was to educate or train administrators and local government officers in public engagement. Participants also recognised a disconnection between concerns and actions. This gap between "knowing and doing" would have to be closed, they believed.

Participants suggested a number of specific actions for the Mayor and the Local Authorities in London.

For the Mayor, the emerging priority was for coherent London wide planning, for traffic, for example. The Mayor should play a leading role in developing community participation in governance, holding public debates and instituting new forum for public deliberation. He should also take steps to ensure that Local Authorities heard the publics' voices: training senior planning officials would be a good start. His aim should be to establish the credibility of public participation in planning and other important public decisions.

The Local Authorities, participants believed, should be prepared to take a lead in instilling a sense of the urgency of London's problems. They should be prepared to innovate and to ensure that considerations of sustainability were more closely integrated into planning decisions.

During the workshop, participants gave many vivid illustrations of their experience of trying to implement more sustainable solutions to urban problems, ranging in transportation from the promotion of cycling to reducing the use of cars in commuting, and in dealing with resources from projects designed to promote the uptake of recycling schemes to those for cutting down on waste.

Some important lessons emerged from these experiences. Citizens needed good clear information to help their choices. Local Authorities and the Mayor's Office could do much to improve the quality of information available to Londoners on the environment and the measures they could take to improve liveability. They should develop new ways of making information it widely available, including the use of the internet (Chapter 15). Solutions needed to be simple and easily understood. For example, more recycling would come from better access to better designed, more easily used, recycling facilities.

Participants stressed the importance of 'bottom-up' measures continually throughout the workshop. There was a warning, however, that such an approach should be backed by evidence of its effectiveness and that studies of governance aspects of sustainable development were urgently needed.

It was concluded that, although participants felt that they had made little progress in utilising sustainable development as a practical concept, they had focussed on the aspects of London that they appreciated most and on the immediate barriers to liveability. They had teased out the connections between the different aspects of sustainable development and had suggested immediate actions that the Mayor and the Local Authorities could take to improve 'joined-up' government and better public decisions. Bottom up approaches to governance with much more public involvement and better information clearly seemed the best course, but their effectiveness needed to be established.

SUSTAINABLE BUILDINGS

David Crowhurst

London's steadily increasing population requires additional housing capacity, as well as the more effective use of London's pre-existing housing stock. Housing and all the other types of buildings put great pressures on the environment and are depleting at an alarming rate the global reserves of natural resources that are used in their construction and operation. The objective of making buildings more 'sustainable' is to help mitigate all these environmental effects. The UK Government has particularly focused on those associated with climate change, such as the predicted increase in extreme summer temperatures in London, which may be associated with poor air quality in inversion conditions, and its threat to human health (which will disproportionately affect vulnerable groups of citizens). Although buildings could be adapted to meet this threat by installing air conditioning, this will lead to increased energy consumption and is not sustainable or practical for the whole population. Indeed reduced energy use, especially through increasing efficiency in buildings, is critical to meeting the UK targets for 60% greenhouse gas reduction. This is one element in the total objective of reducing London's disproportionately large 'ecological footprint'. Organisations such as the Building Research Establishment (BRE), universities supported by the Engineering and Physical Sciences Research Council (EPSRC), and innovative local authorities, housing associations, etc, are undertaking major programmes in research and development of sustainable buildings to address these and other issues. This group addressed the issue of 'sustainable buildings'. Several general conclusions emerged.

(i) The definition of 'sustainable buildings' and their requirements depend broadly on two groups of factors, the first of which is increased 'liveability'. This means that buildings should be designed to be flexible as well as being well laid out. They should be of high quality in terms of design, safety and durability, but at the same time affordable and useable, which includes ensuring that they are optimum for people's health, e.g. in respect of indoor air pollution and the adverse effects of climate change. The location of housing is also an essential aspect of its liveability, especially in relation to good infrastructure (e.g. public transport) and being near or being surrounded by 'green space'. In urban areas, well designed neighbourhoods and housing should also provide a sense of security. The second group of factors concerns buildings' use of resources while meeting the needs of their users. New technological solutions and improved building design can lead to more efficient use of energy, water, materials, etc. Another aspect of efficiency and reducing their environmental impact is to ensure that buildings are durable and the materials recyclable, and that they can be maintained at minimum cost and environmental impact.

(ii) The pros and cons of new buildings that are 'sustainable' were reviewed. The advantages of this approach are most obviously the improved performance and efficiency of the buildings in all its functions and impact, namely reduced energy, water use, waste generation, and running costs, and smaller ecological impacts, for example by ensuring the buildings can be recycled and are constructed from recycled materials. They can also be designed to meet aspects of liveability and employment needs (e.g. communications, indoor environment, transport, space design). In some situations there are disadvantages of 'new buildings', especially where they may compromise the heritage value of existing buildings and landscapes, or their sense of place and ecological value. In addition the new construction may involve excess use of physical/finance resources and use of land.

(iii) An alternative option to constructing new buildings is to transform existing buildings so that they become more sustainable. There are many advantages of this approach, notably less use of resources and efficient use of existing built up areas so that the improvements are more realistic and affordable (and often more popular). Older buildings have increased durability and maintain the heritage value and sense of a place. Such buildings may well have lessons from the past about sustainable construction and planning. However there are disadvantages in attempting the restoration/transformation approach because of its limited flexibility and adaptability. In other words because such buildings have been designed for such different living and working conditions, they sometimes cannot be transformed economically or practically. In some areas the quantity of the fabric and infrastructure of housing stock is so poor that it is not worth while reconstructing, and it is better to rebuild completely.

(iv) Even if decisions have been made in principle and if resources are available, there are still significant barriers to construction of sustainable buildings, namely restrictive regulations based on existing legislation, which for example make it difficult to use efficient energy systems, including renewable energy savings and inter connecting them with the national grid. Additionally planning/building controls do not yet allow local authorities to insist on sustainable housing, though sustainability can be a factor in decision making. Lack of understanding of sustainable buildings by the public and by specialists creates negative attitudes (e.g. based on perceptions of higher costs). Conflicts of interest can inhibit the adoption of sustainable strategies. A supporting infrastructure of supplies and advice is necessary before innovative methods become widespread. Market forces can inhibit the adoption of sustainable building techniques, e.g. through non-holistic practices, but equally they can lead to new and cheaper solutions becoming available, e.g. new renewable energy systems.

Increasingly, through public and private actions, more and more incentives are being introduced to speed up and make possible sustainable buildings projects, services and technologies. Some involve public and fiscal planning measures, e.g. carbon tax and thermal efficiency grants for houses. Moreover general peer group and investor pressure in the business world through campaigns for corporate social responsibility and socially responsible investment can help. For those purchasing new buildings or renovating old ones they are creating a market demand for sustainability. In part this is because people want lower running costs (e.g. for greater energy efficiency). But they also want healthier and better environments, as well as contributing to global environment goals.

RE-ENGINEERING THE HOUSEHOLD — HOW CAN WE MAKE RECYCLING AS EASY AS PUTTING DIRTY WASHING IN THE LAUNDRY BASKET?

Simon Reed

INTRODUCTION

London's households each produce around 800 kg (kilograms) of household waste per year. On average across London, currently only about 88 kg — or 11% is recycled annually. The Government has set targets that by 2005 over 25% of household waste should be recycled, with higher targets for 2010 and 2015. To achieve this each household needs to be separating at least 200 kg of their waste for recycling annually.

It is generally agreed that waste recycling is more effectively achieved through providing special rubbish bins at home specifically for recyclable materials to be stored in, separately from ordinary rubbish. The recyclable materials (e.g. paper, cans, plastic bottles, cardboard and green waste from gardens and kitchen vegetables etc.) are then collected separately or in a multi-compartment vehicle, and sent for reprocessing and recycling, while ordinary rubbish is separately collected and disposed by other means.

However, even where such systems are introduced and recycling bins provided at home, participation in the recycling scheme can be low if residents do not use the bins to separate their recyclable materials. This may be due to a number of factors, one being that the recycling boxes and bins are not conveniently accessible in the house.

THE "DIRTY WASHING" EXPERIENCE

The system of separating materials destined to be recycled is analogous to the common practice at home of using a "linen basket" to store dirty linen prior to laundry. Using this approach each household on average produces, stores and launders around 800 kg/household/year of dirty clothes and linen annually — the same weight as the amount of rubbish produced. By contrast however, we are very much more successful at "recycling" dirty linen. In most households virtually 100% is successfully separated for "recycling" by placing it in the linen basket, so that it can be laundered and reused, whereas where kerbside waste recycling schemes are in operation, the amounts of recyclable waste being separated by householders for collection (88 kg/household/year) is ten times less than the amount of dirty washing being separated for "recycling" via the washing machine.

THE CHALLENGE

If most householders are so good at separating all their dirty washing (800 kg/household/year) and putting it in the laundry basket, what needs to happen to re-engineer the household to enable householders to separate up to 400 kg/household/year of their recyclables, so this amount can be collected for recycling, enabling local and national recycling targets to be met?

THE SOLUTION — "RE-ENGINEERING THE HOUSEHOLD"

We considered the challenge and undertook a scoping exercise with the aim of searching for solutions. The group identified two factors that together significantly affect the effectiveness of a kerbside recycling scheme.

The Design, Ease of Use and Accessibility of Recycling Containers in Homes

The group compared how easy it was to store dirty linen at home compared to storing recyclables and concluded that a number of factors associated with *recycling containers* in homes needed improvement.

(1) Their size and how to identify the recyclable material that each bin or container was intended for (i.e. waste recyclable paper, rather than for plastic bottles and metal cans). A system of colour coding might help.

(2) Bin location. Bins located outside the back door are not as easy to use as (smaller) bins/baskets in each room.

(3) The design of kitchen bins:
 - The need for multi-compartmentalised bins — to enable the separate storage of at least 3 waste/recyclable streams (i.e. dry recyclables (cans, bottles etc); green vegetable waste and ordinary non recyclable rubbish).
 - The need for widely available modular waste and recyclables storage solutions for kitchens in the form of standard size kitchen cabinets with slide-out colour coded receptacles for the main types of recyclables to be separately stored.

Changing Attitudes and Behaviour to Recycling Your Rubbish

The group identified two major barriers to high participation in recycling:

(1) Taking responsibility for your rubbish. Most people are unaware of what happens to their waste. There is a commonly held "out of sight — out of mind" attitude to rubbish. Many people have no knowledge and are ill-informed about what happens to rubbish once the bin is emptied. Few have any understanding of the importance of "closing the loop" so that valuable recyclable resources are re-processed into raw materials and can be remade into new products. Promoting "green procurement" — the purchasing of products made from recyclable materials will assist in this aim.

(2) Providing advice about how to reduce waste production and recycle what is unavoidably produced. Even where people are concerned about the waste they do produce, many are unaware of the options open to them to reduce waste or how to improve their own recycling performance.

The group concluded that a high impact campaign on waste awareness is required. This should provide information about rubbish and what happens to it once the bin is emptied. It should also assist people in learning how to avoid producing waste and how to most effectively recycle. This campaign would need to be reinforced and supported by local campaigns, information centres and other shared resources. More information about the opportunities for "buying recycled" will reinforce this aim.

CONCLUSIONS

The group concluded that there are three main actions required to increase the effectiveness of kerbside recycling schemes.

(1) The designs of rubbish and recycling bins for use in homes need to be radically changed, to enable their deployment and use to become more effective.

(2) These needs to be much more information and support provided to people who are being asked to recycle their waste, so that people's attitude and willingness to recycle changes.

(3) The important environmental and economic benefits of green purchasing should be publicised and promoted to all.

ECOLOGY OF LONDON'S WATERWAYS

Dave Webb

INTRODUCTION

A wide variety of aquatic habitats occur in London, including ponds, lakes, rivers and streams. Each of these habitats has their own specific issues.

It was agreed that the group would primarily address issues relating to large linear water bodies such as canals and rivers, as they represent a significant proportion of the aquatic resource within London. The range of organisations with an interest in rivers and canals is also greater than those associated with still waters.

ISSUES

The overriding issue is that despite significant improvements over the last twenty years, many waterways are still not reaching their ecological potential. This may be because of impoverished water quality due to surface water run-off, habitat loss due to over-engineered channels or due to catchment changes such as the loss of floodplain and other associated wetlands. The result of this is a loss of habitat and associated biodiversity, and a reduced ability of the system to moderate the impacts of storm flows and 'diffuse' pollution (e.g. excess fertilisers and insecticides from farms and gardens).

The implications of this are that the potential for the watercourses to provide diverse habitats and ecological corridors, and the environment for people to interact with the wildlife, has been diminished. If biological diversity is lost then a waterway can become dominated by nuisance species such as midges, to the detriment of local residents. Without the local interest and understanding of what needs to be done, the waterway will inevitably become subject to further abuse, and ultimately become a hostile environment for both people and wildlife.

The group was able to identify the following broad factors that have led to this deterioration. They are complex and often a response to local pressures. Many waterways have been degraded to suit a specific need, for example to maximise the land area that can be developed, or to channel floodwater away as quickly as possible. In doing so, the potential for the watercourse to provide other benefits is ignored. This can be a due to a lack of awareness of ecological benefits or the opportunities to restore them when they have been lost. This is further compounded by developments being driven by single issues and a lack of effective consultation. Consideration of catchment issues, such as diffuse pollution and increased run-off, is generally poor, and the wider consequences of misguided developments and inaction are not fully understood.

THE WAY FORWARD

It was agreed that no one organisation is in a position to resolve all the problems of the water environment, and that partnerships have to be established through a wide constituency of organisations. There needs to be a change in how these organisations work, especially the collaboration with potential partners beyond their own remit and geographical boundaries. There needs to be a shared vision of the future for our waterways. These links and partnerships should not only address established ecological issues but also an integrated approach to call the developments affecting the waterways.

The numerous partnerships currently in existence are not always as effective as they could be, in dealing with long standing problems. Innovative solutions may need to be adopted; the use of facilitators can be an effective way of overcoming problems when projects stall involving several organisations.

It was agreed that to resolve local problems local solutions needed to be developed, and as such it is essential that the community is involved. This would be in the form of enabling community participation, ensuring that the input is effective in informing decision makers and demonstrating the benefits of local input.

Community liaison should be long-term and not just for the duration of a specific project. Experience with ecological projects shows that the right people with the necessary skills are doing the right jobs. There has to be a clear understanding of roles and responsibilities, but there must also be leadership from professional groups to raise realistic aspirations, and highlight the opportunities for improvements.

Limited finance often diminishes local enthusiasm to improve the environment. By highlighting the importance of the environment in regard to health and social cohesion, new partnerships and funding sources can be established.

Improvements to ecology need to be secured through the development control system, such as planning gain. It is essential that funds provided for long term maintenance. The development of partnerships involving ecological organisations needs a long term approach, but grants or contributions are often dependant on yearly bids which can result in uncertainty. The development of environmental trusts to deal with longer term issues was considered to be beneficial.

Technological solutions are available to deal with issues such as improving run-off. But in most instances, such solutions are only incorporated into new developments. Retrofitting of these technologies is also possible and would be beneficial. However benefits need to be demonstrated grants can become available.

Ultimately, to get effective and diverse waterway corridors space is needed. Space to accommodate different roles of the waterway, from recreation, through to education and nature conservation. The ecology is itself dependant on space, and the health of a system is dependant on adjacent wetland, buffering the worst effects of storm flows and diffuse pollution, as well as being unique and valuable habitats in their own right. The value of a waterway, which is only a corridor, will diminish in time. Waterways should be seen as a necklace, where the water leads from one feature of interest to another, and not just simply a conduit to move water from A to B as quickly as possible.

CHANGES IN AVIAN BIODIVERSITY, INCLUDING SPARROWS

Jan Hewlett and Keith Noble

Maps and figures are now available to illustrate the distribution of selected bird species in the London area, and to highlight changes over the past thirty years. These explain links between habitats and, the causes of changes. They also suggest practical ways to conserve the capital's wildlife.

National population trends of breeding birds are one of the fifteen 'headline' indicators, updated annually, which the UK Government uses to measure the quality of life. By choosing birds among more familiar economic and social indicators, the Government recognises the maintenance of biodiversity as a measure of sustainability. Because birds are widespread and diverse, and the data on their distribution are very good, these can be used as 'barometers' of change in the wider environment.

A new Atlas of Breeding Birds of the London Area was published by the London Natural History Society (LNHS) in 2002. This is largely based on surveys made between 1988 and 1994, originally as part of a National Atlas survey, and it enables comparisons with a previous London Atlas produced from records for 1968 to 1972. The maps show the status of each species in three categories — showing evidence of breeding, or presence or absence — in each of 856 tetrads in the LNHS recording area, a circle of radius 20 miles centred on St. Paul's Cathedral.

Jan Hewlett, Editor of the new London Atlas, compared changes between the two atlas surveys for birds of various habitats. Water birds, for example the great crested grebe, grey heron and kingfisher, appeared to be generally doing well, benefiting from a wealth of suitable habitats, from efforts to clean up rivers, and from mild winters. Many farmland birds, for example skylark, grey partridge, and barn owl seem to be in decline, reflecting national trends and intensive agricultural practices.

A selection of other birds showed differing fortunes. At the time of the first Atlas, sparrowhawks were very rare and just starting to recover after the banning of organochloride pesticides which seriously affected them and other birds of prey. Now they are widespread where suitable habitat occurs, even right into the centre of London. Magpies have spread, and amongst woodland birds great spotted woodpeckers and long-tailed tits seem to have extended in range. Nightingales are now scarce, although they have recently begun to colonise some scrubby habitats around gravel pits.

Information from mapped surveys like the London Atlas is complemented by population studies such as the Breeding Bird Survey (BBS), and its predecessor, the Common Birds Census. These were organised by the British Trust for Ornithology, the joint Nature Conservation Committee and the Royal Society for the Protection of Birds. In London BBS sites, blackbirds, starlings and mistle thrushes have declined, but great tits, robins and chaffinches have increased. The annual London Bird Reports and regular newsletters keep the picture up to date.

The house sparrow is especially identified with London, and has been regarded as a very familiar and common bird. Now most people realise that there are not as many as there used to be, and in 2002 it became a Red List Species of high conservation concern in the 'The Population Status of Birds in the UK 2002–2007'. This review by the leading governmental and non-governmental conservation organisations placed the house sparrow on the Red List because its population declined by more than 50% in the past 25 years.

The species is also treated as a special case in the London Biodiversity Action Plan, and was the subject of a popular survey 'Where have all our sparrows gone?' in summer 2002. We showed two preliminary maps, drawn from more than 9 000 completed survey forms. These clearly show that sparrows are particularly scarce in the centre of London, supporting observations that they no longer nest in Kensington Gardens, Hyde and St. James's Parks. There was discussion of possible causes of decline. In the first half of the last century, as horses gave way to motor vehicles, sparrows lost a source of food from the horses' nosebags and droppings. But this effect cannot account for the recent dramatic decline. A shortage of insect food for nestlings, perhaps as a result of air pollution, has been put forward. Other suggestions include increased predation by sparrowhawks, magpies, and cats; development of brownfield sites with their 'weeds' and insects; disease; and a shortage of nest sites in modern styles of roofs.

A report of the survey was produced in January 2003, but further studies will be needed to solve the puzzle of why a species which seemed so capable of living with people in cities is disappearing fast from many parts of London.

In conclusion, birds are among the best studied and most appreciated of our wild animals. In shaping the future of London, we must consider them for their own sakes, and for what they may tell us about the quality of the environment which we share with them.

CLIMATE CHANGE AND LONDON

Tim Reeder

The break out group looked at climate change in terms of Adaptation — the need to adapt to the changes that climate change will bring — and Mitigation — the need to reduce the emissions of greenhouse gases, which are thought to be causing climate change, in both the short and long term.

ADAPTATION

The following points were debated and highlighted as issues of significant concern:

(1) The risk to the Estuary and the Thames riverside was identified as perhaps the biggest threat to London. It was seen as a priority to investigate the threat to the effectiveness of the Thames tidal defences from the probability of rising sea levels and more frequent river floods. In

addition the possible, though less certain, increase in storminess could lead to increased wave heights and associated erosion rates on key parts of the Thames estuary. The Environment Agency is starting to address this threat through its new flood risk management project, which is concentrating on future requirements for flood risk management in the Thames estuary over the next 100 years and more.

In addition the possibility of increased rainfall was identified as being a risk to urban flooding. Current surface and storm sewerage systems were built assuming maximum short term precipitation rates that could well be exceeded given climate change.

(2) The likelihood of hotter, drier summers could threaten the already highly stressed water supply/demand balance in the Thames Valley on which London is reliant for its water supply. It is likely that greater demand management measures and perhaps major water supply infrastructure investment, for example on new reservoirs, will be required to meet this challenge.

(3) It was thought that the trend of increasing temperatures will exacerbate the existing Heat Island Effect during summer in London. This will exacerbate health problems and make working conditions difficult — increasing the demand for air conditioning, which in turn will increase energy use and associated greenhouse gas emissions. Associated with the Heat Island Effect is poorer air quality which causes further health problems for London's citizens. By contrast higher winter temperatures might reduce the number of hyperthermia related deaths.

However, it is likely that London will still be considerably cooler than southern Europe where temperatures could become very extreme and unpleasant and where currently millions of people holiday each year. Thus London's tourist industry could benefit in the future as more people chose to holiday here.

(4) The likely increase in storms and floods due to climate change will have implications for infrastructure such as power lines. In addition more frequent extreme weather events will put greater strain and emphasis on the need for well prepared emergency planning and associated services.

(5) Climate change will have serious implications for the built environment. For example extreme summer temperatures would require greater shading in buildings. To minimise fossil fuel use the heating and energy requirements of buildings will increasingly need to be from

alternative renewable sources such as solar power. Transport systems such as the Underground will need to adapt to higher and possibly uncomfortable temperatures in summer. Some workshop members thought we might experience "building failure" due to causes such as high temperatures and high wind speeds. London could benefit hugely by developing and investing in environmental and building technologies to adapt to climate change.

(6) Biodiversity and natural systems might alter given a changing environment. Possibly we should move towards conserving potential habitats rather than trying to preserve species that may well need to move/adapt given differing conditions. The issue of invasive species and whether they are sustainable was discussed.

(7) London's position as a world finance and insurance centre will expose it to the risks posed at a global level by the impacts of climate change. Its exposure and resilience needs to be investigated. London could be influential in setting in train financial instruments to manage risks. In addition it could help lead on economic incentives to reduce greenhouse gas emissions.

SHORT TERM MITIGATION

(1) The profile of energy efficiency needs to be raised, especially in buildings. As well as supporting campaigns such as those of the Energy Savings Trust, it would be sensible, and act as an incentive, to have demonstration projects. The new Bedzed development in south London is the best current example. Encouraging more such examples is essential.

(2) It was thought that traffic free or low energy days are worthwhile, provided people understand why energy reduction is required. Demonstrations are needed to show how we can change our life style to reduce the current unsustainable level of greenhouse gas emissions.

(3) Another approach is to promote labelling of products with the energy efficiency involved in their manufacture. There was some debate as to how effective this is in the absence of related economic incentives.

(4) The issue of better standards for new buildings in terms of energy efficiency and embedded renewable energy generation is seen as critical. These are proposals for supplementary planning guidance to encourage

this in the Mayor's draft energy strategy. In addition London could lead in calling for major improvements in the building regulations.

LONG TERM MITIGATION

(1) In the longer term London could be very influential in encouraging the planned global reduction of greenhouse gases (a reduction in emissions of 60% over the next 50 years), particularly if this message was taken on board fully by London's financial and investment community. These targets would relate to local targets. The London Sustainability Commission are working to attain these longer term major reductions. The British Government's energy white paper (February 2003) has adopted this long term goal.

(2) There was debate as to how we would reach these long term reductions. The balance between altruism and the need for economic incentives was debated. The conclusion was that both have a role.

(3) There is an urgent need for research and development into technologies that could deliver these long term reductions. London is well placed to lead this and institutions such as Imperial College are already making headway. The Building Research Establishment has a major programme on sustainable buildings, which was presented to the conference.

(4) It was thought fundamental for real political will to drive this longer term issue forward (especially since the withdrawal of the USA from the Kyoto Protocol). The population as a whole need to be educated and convinced of the urgent need to make the necessary major changes in economy and lifestyles. London could help lead on this. The Mayor is in a good position to help convince the average Londoner of the need to change. He has a track record of being straightforward with his electorate and his style could well be useful in getting the message over. The group also debated how celebrities and sports stars could play a leadership role in helping to gain ownership and change minds and attitudes, David Beckham for example!

London, along with New York and Tokyo, is the world's major financial centre. The Mayor of London and other lead figures of its institutions could lead in convincing important global networks and institutions to support change.

MONITORING THE ATMOSPHERIC ENVIRONMENT OF LONDON

Roy Colvile

This breakout group considered in its widest sense, including the state of London's atmospheric environment that people experience in the streets, buildings and open spaces. The meteorological parameters are changing and likely to change in the future, as well as the form and level of air pollution. Some of the factors that make the environment of London unpleasant and unhealthy are shown in Box 1. (These resulted from the initial 'brainstorm'.) Half the group considered the policy of doing everything possible to improve the environment of London, while the other half of the group considered the policy of simply leaving London to develop economically without worrying about its climate.

The pro-intervention group pointed out that current trends in the deterioration of London's environment cannot be sustained much longer. Gridlock, for example, is uneconomic and can kill an urban centre. The costs of health impacts of poor climate are high, including the effects of the stress of living and working in London. London's activities also have an unacceptably and unsustainably large global impact. Radical policies to address these problems might change anything that gets in the way of improving London's environment. These could perhaps provide win-win solutions that, far from being expensive, would bring both economic as well as environmental benefits.

There are also compelling arguments for laissez faire policies based on live and let live... and making money. Individual freedom to seek

---

Box 1 What makes the environment of London unpleasant and unhealthy?

Congested traffic	Rotting rubbish	Litter
Sewers	Grime	Hot
Noisy	Hurry	Humid
Lack of greenery	Drainage	Busy
Can see effects	Tube environment	Windy
Urban layout	Smells	Can't go to clean places
Grey sky	Feels dirty	Dust
Puddles	No sky	Money matters

information and to take the most effective and economic action is preserved when environmental improvements are not imposed from above. Perhaps this could lead to market forces allowing a happy equilibrium to be established. Market forces may have a further benefit. London is a stressful place, but that stress leads to survival of the fittest and helps us to maintain our competitive position as a leading world city — a kind of Darwinian monetarism?

The group then considered what role monitoring can play in reconciling these two polarised views. It became clear why monitoring the environment of London has an important role to play. Both sides of the argument use data from monitoring to support their points. Both sides of the argument need monitoring to help strengthen the economic, environmental, or social pillars of sustainable development (depending on which pillar is their favourite one!). Furthermore, the use of monitored data helps move the debate from speculation and assertion about critical and ongoing evaluations of London's environment. This is essential for a well-informed resolution of policy differences.

The group was surprised by the large number of factors listed in Box 1 that determine London's environment. We suspected that most of these natural and social parameters are in fact being monitored by a variety of organisations with their various particular interests. Finding the data however, is far from easy, as it is scattered widely. It was concluded that a useful output from the LEAF conference and its successors would be a meta-database indicating what data exists on monitoring the climate and environment of London. By covering as wide a scope as possible, such a meta-database would help all sides of the environmental debate in London. This would help research and would enable communities, government and business reach better informed decisions to benefit London's whole environment.

MONITORING AND REDUCING NOISE

Jenny Stoker

INTRODUCTION

According to a recent survey (Annual London Survey 2001), 26% of people who live in Greater London consider noise to be a major problem. It would seem, therefore, that London needs to deal with noise. In another survey (The 1999/2000 National Survey of Attitudes to Environmental Noise) where people were asked to specify the sources of noise by which they were 'moderately', 'very' or 'extremely' bothered, annoyed or disturbed, the four most common noise sources were:

- road traffic 35% (22%)
- neighbours and/or other people nearby 28% (19%)
- building, construction, demolition, renovation or road works 13% (7%) and
- aircraft/airports/airfields 6% (7%)

(with the percentages in brackets being values for the UK as a whole). Here, the road traffic, and noise related to aircraft and airports are classified as *ambient noise* sources, whereas 'noisy neighbours' and building works are considered as *noise nuisance*.

The recent noise initiatives within the UK regarding ambient noise have been motivated by the EU directive (2000). The Government has published a consultation paper (2001) 'Towards an Ambient Noise Strategy', and the Mayor of London has written a 'Draft Ambient Noise Strategy' (2002). Some of the London boroughs have published their own noise strategies, for example Camden (2002). In contrast to the other documents mentioned above, the borough noise strategies have to address noise nuisance in addition to ambient noise.

However, despite all these strategies, UK national noise limits have not yet been set. The World Health Organisation (2000) recommends that the day-time outdoor living area noise levels should not exceed $55\,\mathrm{dB}\,L_{Aeq}$ (L_{Aeq} refers to the 'equivalent' average sound level measured using the A-weighting which is most sensitive to speech intelligibility frequencies of the human ear) and the night-time value at the outside facades of living areas should not exceed $45\,\mathrm{dB}\,L_{Aeq}$ (so that people can sleep with their windows open). Applying these levels, 55% of the population of England and Wales are exposed to levels exceeding this day-time value, and 68% are exposed to the night-time value (when the day is taken to be from 0700–2300, and night to be 2300–0700). As noise levels in London tend to be higher than in other parts of the UK, corresponding statistics in London will be higher.

NOISE MAPPING

In a Rural White paper (2000) the Government proposed to develop a national noise strategy for England. They have set aside £13 million to fund this initiative that, in the first stage, will involve the noise mapping of all large agglomerations (above 250 000 inhabitants), major road and rail links. It is intended that this work will be finished by 2004. A noise map of London will therefore be produced in the next couple of years. Noise mapping is a useful tool in integrated transport policy. It gives a good picture, and indicates hotspots. However, one drawback of this recent initiative may be that the noise mapping work is being done by consultants, as opposed to local authorities. This is in contrast to the recent Air Quality Review and Assessment procedure (2000), where local authorities have taken an active role in the production of 'base case' air pollution maps, and in the subsequent investigation into the effect of air quality action plans. Therefore, in order for local authorities to be able investigate the effects of any proposed noise strategies, they must given full access to the source data and training in the relevant noise mapping software.

SOME SOLUTIONS FOR LONDON

There are a number of ways to reduce ambient noise levels in London. For example:

- the Mayor's congestion charging should reduce noise levels, as traffic will travel more smoothly, and there will be less 'stop-start' driving
- improvement of the road surfaces significantly reduces noise levels, and improvement of railway track quality and maintenance on both National Rail and the Underground will have a similar effect
- reduction of the number of heavy goods vehicles travelling within the London area — perhaps by use the rail freight system instead
- use of quieter buses and other vehicles, for example electric cars, fuel cell buses and hybrid-electric buses, and
- imposing a night aircraft ban over London.

Some suggestions for reducing noise nuisance are:

- improvement of sound insulation within properties — this problem should be addressed in the new building regulations (1984)
- advice and guidance should be given to businesses regarding ways they can reduce their noise output, in particular late at night and in areas of high population density
- local authorities should have sufficient resources to ensure that noise levels are monitored and/or recorded and appropriate action taken

- the public should be educated regarding acceptable levels of noise, and
- improved planning policy guidelines to ensure more consistent decision making.

SOME ADDITIONAL IDEAS

Monitoring — A national noise monitoring network should be set up which would include a number of noise monitors within London, and would be similar to the UK Air Quality monitoring network. These monitoring data should be used to validate the noise maps.

Noise Source Apportionment — Total noise levels in London are a combination of noise from different sources. Also, the contribution from each source may have a diurnal and weekly variation. In order to reduce noise levels, it is important to know where the noise comes from.

Parks and Open Spaces — Results of the Noise Incidence Survey (2002) show that noisy areas are getting slightly quieter but quiet areas are getting slightly noisier. This means that in London, the areas that have been seen as quiet spaces, such as all the parks, are subject to a creeping noise effect. This needs to be addressed.

Aircraft Noise — Airports are predicted to expand massively over the next few decades. The Government is currently undertaking the biggest public consultation exercise the UK has ever seen. The siting of new runways and continued expansion will lead to significant impacts of noise from aircraft as well as noise impact due to increased surface access requirements and other major infrastructure requirements that will also be required. Increased demand for commercial land usage close to major airports and the knock on effect of increased demand for residential dwellings will also have significant impacts on the local noise climate. It is important that decision makers take on board all the associated effects of these issues and either ensure that appropriate amelioration measures are adopted, or where noise impact is very severe, significant noise sources are separated from communities.

Research — Though some effects of noise are understood, there are still large gaps in knowledge relating to the effects of noise and there is a need to conduct more longitudinal studies to further investigate the effects of noise on health.

REFERENCES

Annual London Survey (2001) Londoner's views on life in the capital. MORI for Greater London Authority.

The 1999/2000 National Survey of Attitudes to Environmental Noise (2002) Volumes 1–5, BRE Client Report Nos. 205215f-205219f.

Directive of the European Parliament and of the Council relating to the Assessment and Management of Environmental noise, EU COM (2000) **468**.

Towards a National Ambient Noise Strategy (2001) A Consultation paper from the Air and Environmental Quality Division, Department for Environment, Food and Rural Affairs, London.

The Mayor's Draft London Ambient Noise Strategy (2002) Assembly and Functional Bodies Draft, Greater London Authority.

Noise Strategy (2002) London Borough of Camden.

Guidelines for Community Noise, World Health Organisation, Geneva (2000) http://www.who.int/peh/noise/noiseindex.html.

Our countryside: the future — A fair deal for rural England. Paragraph 9.4. (2000) Command 4909. The Stationery Office.

Review and Assessment: LAQM TG1-4(00) (2000) Department of the Environment, Transport and the Regions, Scottish Executive, National Assembly for Wales.

Building Regulations, Building Act 1984, Approved Document E: Airborne and Impact Sound.

The UK National Air Quality Information Archive, http://www.airquality.co.uk/archive/index.php.

The National Noise Incidence Study 2000/2001 (United Kingdom) (2002) Volumes 1–2, BRE Client Report Nos. 206344f-206345f.

LAGOON FOR LONDON

Richard Layard

PROPOSAL

Imagine London as the Venice of the North, with a river which was generally high — beautiful to the eye and drawing people to it. Cafes, galleries and meeting places would spring up along its banks, and riverside housing would rise in value. The river would become the centre of the activities. It would be one of the spectacles of the world.

Is it possible? The answer (with reservations) is Yes. We need simply to raise the Woolwich Barrier and thus hold up the water level in the capital. This would create a lagoon through the city, right up to Richmond.

A hundred years ago the citizens of Richmond had a similar idea for their own stretch of the Thames. They wanted a beautiful river that was always full of water. So they built a barrier below Richmond that held in the water above it and created the beautiful tranquil stretch of Richmond Reach.

For the rest of London we do not need to build a barrier; it already exists. The question is whether we want to use it to make the rest of the Thames as beautiful as it is at Richmond. I want to discuss the pros and cons of doing this, and above all the case for a systematic enquiry.

BENEFITS

A lagoon in London would be a thing of beauty, a place where people wanted to be — to sit, to walk and to gaze. Tourism would thrive. But more important, millions of Londoners who never go near the river would consider it the place to be. As Parisians flock to the Seine, or Bostonians to the Back Bay, so Londoners would flock to the Thames. The city's focus would change and eventually the gain in economic value could run into billions of pounds.

The view of the river from the land would be transformed, but so would the view of the land from the river — making river trips a joy at all times of the day. We could also have a much better system of water transport, similar to the vaporetti in Venice. As congestion grows on land, it will be ever more important to use the opportunity which the river provides.

Other forms of river use could also be developed, subject to suitable controls. On parts of the river sailing, canoeing and boating could become spectacular events.

THE POSSIBILITIES

However there are also significant problems. That is why we should consider different variants of lagoon. Each variant should be submitted to rigorous appraisal of its costs and benefits. The authorities should undertake such a study and publish the results and their proposals for action.

The most ambitious variant is to hold the river at continuous high tide. This would produce the most beautiful effect, and thus yield the largest benefits. But it would also raise the largest problems.

A more limited idea would be to ensure that the water in the river never goes below its level at half-tide. That is the system at Richmond. Under this arrangement there is a normal high tide (with no barrier). But when the tide goes below half-tide, the barrier is put in place. So the water level never goes below half-tide, and for half the time it stays at that level — beautiful and still. The mudflats never appear. To see how half-tide would look in the more built-up environment of central London, one can go to St. Katharine's Dock, or other parts of Dockland. The impression is very attractive.

Another variant would be to allow normal low tides when these would happen in the middle of the night. All these possibilities should be explored.

SO WHAT ARE THE MAIN PROBLEMS?

Sewage — At present raw sewage enters the Thames whenever there is a flash flood of rain. This is because in London rainwater and sewage go down the same pipes and the overflow from the pipes goes into the river.

Thames Water have plans to eliminate this arrangement over time. These changes need to be speeded up. But, until they are completed, one might have to be content with a half-tide lagoon. Though this only requires that the barrier at Woolwich be half-raised for half of the time. For the rest of the time, the barrier would be completely open. But, even with this, it might be necessary to have days when the tide was allowed to go right out. All this requires study.

Seepage — Another problem with a fuller river is a greater seepage of water into the sub-soil. The level of London's water-table is already rising, damaging the foundations of some buildings and causing water to seep into the Underground. The existing problem will of course have to be addressed anyway and a 1989 study suggested it could be quite effectively handled by drilling boreholes at a cost of under £16 million.* The extra cost if the Thames were fuller could only be discovered by further study.

Silting — A fuller and stiller river could also deposit more silt. When a barrier was first proposed at Richmond, the Thames Conservancy opposed it on these grounds and they managed to delay its construction for decades. Their objections proved in the event to be largely unfounded. Thus, while there could certainly be some extra cost of dredging, this is unlikely to be the central issue.

---

*Discussed in GARDIT Strategy Proposal, Controlling London's Rising Groundwater, March 1999, p. 9.

Shipping — This also is unlikely to be the central issue. At present the river is used to ship waste down to landfills in Essex, and to bring up building aggregates to wharves in West London. With a half-tide solution, this traffic could probably continue. However, the issue of waste may disappear in 10 years when the Essex landfill is complete.

Sea-life — The problems discussed so far are essentially those of cost. But then there are the more intangible issues of ecology. The mudflats above Woolwich would be permanently covered. Birds which now feed on these mudflats would have to feed a few miles further down the river. New patterns of life would develop in London's river.

For some people the old ecological patterns are so important that they override most other considerations. Policy-makers should be sensitive to these feelings, but one does have to recognise that without ecological change we should never have had a city in London, and certainly no Embankment.

CONCLUSION

I conclude that we need a serious study of all the possibilities — their costs and benefits. There has been no substantial look at this issue since the 1960s. Since then people have come to value the environment of their cities a great deal more than they did. That is why we should dare to imagine a very different London.

To use the Woolwich Barrier in this way would require primary legislation. So did the Richmond Barrier. Its opponents fought it for 30 years, and eventually it passed the commons by only 106 to 100. In Cardiff Bay and the Tees estuary we have spent hundreds of millions on new structures aimed to create a watery delight. In London we already have a structure capable of producing one. Let us at least systematically examine how we might use it.

DISCUSSION

The group was divided on the merits of the proposal. There were two main objections.

The first was based on the desire to preserve the natural tidal environment, including the availability of the foreshore at low tide to birds and people. It was agreed that the view of the river and its accessibility would be better if the river defences were stepped instead of being vertical. It was also argued that the tidal Thames is an important breeding ground for North Sea fish — though relevant statistics were not available. The second objection was that the barrier, though it could indeed face inward, might become weakened thereby. Since the flooding situation may require action before 2020, supporters of the lagoon argued that the best opportunity to achieve it might be in the context of a restructured system of flood defences.

A member of the River Thames Society reported that it has long favoured a tideless Thames.

YOUNG PEOPLE'S PERSPECTIVES ON LONDON

Children of Eastbury Secondary School
(and Carolyn Stephens)

We attended the LEAF (London's Environment and Future) conference on 17th and 18th September 2002. The meeting was organised by a committee of people from universities, government institutions and agencies. Its intentions were to help bring people together who were working on research and everyday issues, such as medicine, technology, floods, planning, transport and other such matters. Alternative methods to problems were suggested, and new methods to problems were put forward alongside the highlighting of important issues.

Our group (Eastbury School) got together with some primary school students from Swaffield School and put our opinions across to the panel and audience, as we are the people who face these problems in day to day life. To do this, we signed up for one of the breakout groups, for which we got into smaller groups and brainstormed or ideas, resulting in a 5 minute presentation (per group) on how to make London and its environment a healthier and safer place.

Over the two days, we heard presentations from Lord Rogers, Michael Meacher (Minister for the Environment), and the Mayor of London, Ken Livingstone. Together, they talked about working towards a more sustainable future for London. A reception was held at the Welcome Wing of the Science Museum, which put on displays and an exhibition on climate change. Verbal and song presentations were also put on, along with slides showing London's past.

Iram Zahid, Year 10 Eastbury Comprehensive School

Author Biographies

HELEN APSIMON

Helen ApSimon is Professor of Air Pollution Studies at Imperial College London, heading a diverse team of atmospheric scientists in the Department of Environmental Science and Technology. She has particular interests in urban air pollution and its control; and in transboundary air pollution and integrated assessment (including work under the Convention on Long-Range Transboundary Air Pollution). She is a member of the UK Air Quality Expert Group, and chairs the APRIL (Air Pollution Research in London) network.

DAVID BANISTER

David Banister is Professor of Transport Planning at University College London and has been Research Fellow at the Warren Centre in the University of Sydney (2001–2002) and was Visiting VSB Professor at the Tinbergen Institute in Amsterdam (1994–1997). He has an international reputation in linking transport analysis to the wider issues of urban development and sustainability. He is the author and editor of 17 research books, including **Sustainable Urban Development and Transport** (2004), and is editor of the international journal **Transport Reviews** and joint editor of the journal **Built Environment**.

MICHAEL BATTY

Michael Batty (m.batty@ucl.ac.uk) is Professor of Spatial Analysis and Planning, and Director of the Centre for Advanced Spatial Analysis (CASA) at University College London (UCL). He holds a joint appointment between the Bartlett School of Architecture and the Department of Geography. His books range from **Urban Modelling** (Cambridge UP 1976) to **Fractal Cities** (Academic Press 1994). He is editor of **Environment and Planning B** and a Fellow of the British Academy. He has recently co-edited **Advanced Spatial Analysis** with Paul Longley (http://www.casabook.com).

SUSAN BATTY

Susan Batty (susan.batty@ucl.ac.uk) is Senior Lecturer in Planning and Year Tutor of the MSc/Diploma Town and Country Planning at the Bartlett School of Planning, University College London. From 1996–1998, she was a Specialist Assessor for the HEFCE (Higher Education Funding Council for England) Teaching Quality Assessment Programme in the area of Town Planning. Her research interests focus on institutional analysis, specifically exploring the interface between land use planning agencies and the public. She coedited (with Antonia Layard and Simin Davoudi) **Planning for a Sustainable Future**, Routledge/Spon (2001).

ELSPETH DUXBURY

Elspeth Duxbury is a partner in the consultancy Intelligent Space Partnership and co-founded the firm in Spring 2000. Intelligent Space Partnership is based in Central London and specialises in the modelling and evaluation of how people move around buildings and urban spaces, providing evidence to support user-centred design. Elspeth trained initially in Chemistry graduating in 1996 then completed a Masters in Transport Planning at the University of Newcastle in 1997. Prior to starting Intelligent Space, Elspeth undertook doctoral research at University College London, working at the Centre for Advanced Spatial Analysis also at UCL.

HELEN EVANS

Helen Evans graduated in 1992 with a master's degree in civil engineering. She subsequently worked for the Centre for Environmental Health

Engineering at Surrey University assisting in developing appropriate water treatment systems for developing world countries, working both in Latin America and in the Caribbean.

Subsequent to this she under took an Engineering Doctorate in Environmental Technology run jointly by Surrey and Brunel Universities. Her research was sponsored by Thames Water Utilities Limited, and assisted in developing an understanding of the characteristics of biological water filtration systems, concentrating on pathogen removal.

She has since worked with Thames Water, working in their research and development group, focusing primarily on pesticide removal from drinking water, then within the corporate Strategy group where she reviewed future technology and its potential impacts on water treatment. She is now responsible for formulating the research programme for Thames Water's R&D group.

STEVE EVANS

Steve Evans (stephen.evans@ucl.ac.uk) is a Research Fellow in GIS at the Centre for Advanced Spatial Analysis (CASA) at University College London where he works on 3-dimensional computer models of parts of central London and Environmentally focused GIS such as the recent EU funded PROPOLIS project focusing on the "City of Tomorrow". He is also a Director of Planet Visualisations Ltd. Previously he worked on GIS and Environmental Data Management for British Antarctic Survey.

JONATHAN GLANCEY

Jonathan Glancey (jonathan.glancey@guardian.co.uk) is Architecture and Design correspondent of the Guardian. A frequent broadcaster, he is author of The Story of Architecture, Twentieth Century Architecture, The Car: A Social History of the Automobile, and London: Bread and Circuses, an essay concerning political fashions and the follies of contemporary planning, or lack of it, in the capital.

DAVID GOODE

David Goode was recently Head of Environment at the Greater London Authority and is a Visiting Professor in Geography at University College London. He has held senior posts in both central and local government, specialising for the past twenty years on the urban environment. He is a past President of the Institute of Ecology and was formerly Director of the London Ecology Unit. At the GLA, he has been responsible for developing a suite of environmental strategies, including Waste Management, Air Quality, Biodiversity, Noise and Energy. As an ecologist he has argued for a radical approach to the management of cities based on ecological principles, and has aimed to ensure that the strategies for London provide a framework for achieving environmentally sustainable solutions. Professor Goode has written extensively on environmental issues, including an essay on *Cities as the Key to Sustainability*. He gave the Brian Walker Lecture on Environment and Development in 2003.

PAUL HENDERSON

Paul Henderson is an Honorary Professor in the Department of Earth Sciences at University College London. He recently retired from the position of Director of Science at the Natural History Museum, London. He has been active over several years in promoting the roles of taxonomy and systematics in sustainable development. He is also Vice-President of the Geological Society of London. His e-mail address is P.Henderson@btinternet.com.

ANDY HUDSON-SMITH

Andy Hudson-Smith (asmith@geog.geog.ucl.ac.uk) is Research Fellow and Systems Administrator at the Centre for Advanced Spatial Analysis (CASA) at University College London. Trained as an urban planner, he specialises in the development of multimedia, 3D GIS, and computer-aided design for urban and environmental problems. His focus is on public participation and the development of environmental awareness using digital technologies in the community. He is currently working on the Virtual London Project for the Greater London Authority which is part of their e-democracy initiative. He has published many papers in this area and his work is displayed at http://www.onlineplanning.org.uk/.

JULIAN HUNT

Julian Hunt is a Professor of Climate Modelling in the Departments of Space and Climate Physics and Earth Sciences at University College London. He is a Fellow of the Royal Society and Chairman of Cambridge Environmental Research Consultants Ltd. Formerly he was Chief Executive of the Meteorological Office and UK's representative at the World Meteorological Organisation (WMO). He spoke on behalf of the WMO at the UN conferences on Natural Disasters and on the Future of Cities in 1994 and 1996. In his political life in the Labour Party he was a City Councillor in Cambridge in the 1970s and is now in the House of Lords. His e-mail address is jcrh@cpom.ucl.ac.uk.

KEN LIVINGSTONE

Ken Livingstone was elected first Mayor of London as an independent in 2000, following the reformation of the Government of London that year, that also created the Greater London Assembly which advises him. He was a Labour member of the Greater London Council from 1973–1986, being its leader from 1981–1986. He was Labour MP for Brent East 1987–2001. Because of his lifelong interest in natural history he was a member of the Council of the Zoological Society of London and later elected a Fellow. He is author of two books, 'If Voting Changed Anything they'd Abolish it' (1987), Livingstone's Labour (1989).

MICHAEL MARMOT

Michael Marmot is Professor of Epidemiology and Public Health, and Director of the International Centre for Health and Society (established in 1994), at University College London. He is Principal Investigator of the Whitehall studies of British civil servants, and Principal Investigator of the English Longitudinal Study of Ageing. Together with the National Centre for Social Research, the International Centre for Health and Society conducts the Health Surveys for England and Scotland. He was a member of the Chief Medical Officer's Committee on Medical Aspects of Food Policy and of the Chief Medical Officer's Working Group on "Our Healthier Nation"; member of the MacArthur Foundation Network on Socio-Economic Status and Health; and member of the Royal Commission on Environmental Pollution. He served on the Scientific Advisory Group of the Independent Inquiry into Health Inequalities chaired by Sir Donald Acheson, which reported in November 1998. He was knighted by HM The Queen in 2000.

JACQUELINE MCGLADE

Jacqueline McGlade became Executive Director of the European Environment Agency in Copenhagen in 2003; she is on leave from her post as Professor in

Environmental Informatics in the Department of Mathematics at University College London. Until 2003, she was a Board member of the Environment Agency of England and Wales with responsibility for Thames Region, navigation and science. Formerly she was Director of the NERC Centre for Coastal and Marine Sciences, Professor of Biological Sciences at Warwick, Director of Theoretical Ecology at the Forschungszentrum Juelich and senior scientist in the federal government of Canada. Her research has focussed on the spatial and nonlinear dynamics of ecosystems, with particular reference to marine resources, climate change and scenario development. In her non-academic life she is a mother of two daughters, director of a software development company and has written and presented a range of radio and television programmes. Her e-mail address is jacqueline.mcglade@eea.eu.int.

MICHAEL MEACHER

Michael Meacher was Minister of State for Environment (1997–2001), where he was responsible for environmental legislation and he represented the UK internationally at Kyoto in 1997, Johannesburg in 2002 and other major environmental negotiations. He is now Chairman of the UK branch of GLOBE (Global Legislation for a Balanced Environment). Formerly he was the Labour Chief Spokesman and Parliamentary Under Secretary in health and social security, employment, overseas development, transport, employment trade. He has been Labour MP for Oldham West (and Royston) since 1970. He was a lecturer in social administration at University of York and LSE. His books include 'Socialism with a Human Face' (1981), 'Diffusing Power: the key to Socialist revival' (1992).

GEMMA METHERELL

Gemma Metherell was educated at Sexey's School, Bruton, England and then graduated from The University of Birmingham (UK) in 2002 with an honours degree in Geography. Her work on Monet and his depiction of weather in his London Series reflects her interest in both Art and Geography. She is now pursuing a career as an Officer in the British Army.

TARYN NIXON

Taryn Nixon is the Managing Director of MoLAS, the Museum of London Archaeology Service, an organisation of around 180 professional archaeologists and heritage specialists. MoLAS and its Specialist Services arm provide consultancy advice and practical services to the commercial property development and heritage sectors, and carry out archaeological fieldwork and research. Taryn is actively concerned with integrating archaeological endeavour with sustainable development. She is a member of the CABE (Commission for Architecture and the Built Environment) Design Review Committee, a member of the NERC Science-based Archaeology strategy group and a former Chair of the professional body the Institute of Field Archaeologists.

DENNIS PARKER AND EDMUND PENNING-ROWSELL

Dennis Parker and Edmund Penning-Rowsell are Professors of Geography at Middlesex University. Both have been researching the socio-economic, organisational and political aspects of floods and related hazards for over thirty years through the Flood Hazard Research Centre at Middlesex University. In 2000 this centre led by Edmund was awarded a Queen's Anniversary Prize for education, training and research. The citation says that 'This world-class centre is at the forefront of efforts to ensure more sustainable protection worldwide for communities at risk from flooding'. Both authors are also Pro Vice-Chancellors at Middlesex University. Their e-mail addresses are: D.Parker@mdx.ac.uk and E.Penning-Rowsell@mdx.ac.uk

RICHARD ROGERS

Richard Rogers is one of the foremost living architects, the recipient of the prestigious RIBA Gold Medal in 1985 and winner of the 1999 Thomas Jefferson Memorial Foundation Medal and the 2000 Praemium Imperiale Prize for Architecture. Richard Rogers was awarded the Légion d'Honneur in 1986, knighted in 1991 and made a life peer in 1996. In 1995 he was the first architect ever invited to give the BBC Reith Lectures — a series entitled 'Cities for a Small Planet' and in 1998 was appointed by the Deputy Prime Minister to chair the Government's Urban Task Force. Most recently he was appointed as Chief Adviser to the Mayor of London on Architecture and Urbanism and also serves as Adviser to the Mayor of Barcelona's Urban Strategies Council. Richard Rogers has also served as Chairman of the Tate Gallery and Deputy Chairman of the Arts Council of Great Britain. He is currently a Trustee of the Museum of Modern Art in New York.

Richard Rogers is best known for such pioneering buildings as the Centre Pompidou, the HQ for Lloyd's of London, the European Court of Human Rights in Strasbourg and the Millennium Dome in London. His practice, founded in 1977, has offices in London, Barcelona and Tokyo, and is now engaged on two major airport projects — Terminal 5 at London's Heathrow Airport and Barajas Airport, Madrid, currently the largest construction site in Europe. Other schemes include high-rise office projects in London, a new law court complex in Antwerp, the National Assembly for Wales in Cardiff, a hotel and conference centre in Barcelona and a new bridge in Glasgow. The practice also has a wealth of experience in urban masterplanning with major schemes in London, Shanghai, Berlin, Florence and Lisbon.

MATTHEW SIMMONS

Matthew Simmons is a graduate of Chemical Engineering and worked with Arup's environment and sustainability form 2000 until 2004.

MAI STAFFORD

Mai Stafford is Senior Research Fellow in the Department of Epidemiology and Public Health at University College London. Her research interests focus on the macro-social determinants of health, especially the relation-ships between social and economic features of the residential environment and health.

JOHN E. THORNES

John E. Thornes is a Reader in Applied Meteorology in the School of Geography, Earth and Environmental Sciences at the University of Birmingham. He runs a final year course in 'Geography and the Visual Arts' and recently published a book on 'John Constable's Skies'. He has always been fascinated by the different approaches artists use to paint the weather.

TONY TRAVERS

Tony Travers is the Director of the Greater London Group at the London School of Economics (LSE), which carries out research into subjects relating to the government and economy of London and the South-East of England. His publications include 'The Politics of London: Governing an Ungovernable City' (with M. Kleinman) (2003) and 'Decentralization London-style: The GLA and London Governance' (2002).

LORNA WALKER

Lorna Walker has recently retired as Director of Ove Arup & Partners with over 25 years experience in environment and sustainability. She is Visiting Professor of Engineering Design for Sustainable Development at the University of Sheffield, and a member of the International Federation of Consulting Engineer (FIDIC)'s Sustainable Development Task Force. In the past she has been a member of Sir Martin Laing's Sustainable Construction Focus Group and Lord Richard Rogers UK Urban Task Force, and a member of the Institute of Civil Engineers Environment and Sustainability Board.

WILLIAM WESTON

William Weston has an extensive background in the management of high profile cultural organisations, such as the West Yorkshire Playhouse in Leeds, and the Royal Shakespeare Company. He and his team at The Royal Parks have played pivotal roles in major national events such as the Golden Jubilee and the State Funeral of the Queen Mother. William is committed to reconnecting people and local communities with the natural world and is a vigorous advocate for the social, educational and health benefits of urban green space.

ROGER WOTTON

Roger Wotton is a Professor of Biology at University College London. He teaches Aquatic Biology (a fusion of Oceanography, Marine Biology and Freshwater Biology) and studies the role of aquatic animals as "ecosystem engineers". He has collaborated on several projects with the Water Industry and has a strong interest in the provision of drinking water in a sustainable World. His e-mail address is r.wotton@ucl.ac.uk.

NICKY GAVRON

Nicky Gavron is the Deputy Mayor of London. She has been at the forefront of environmental and transport issues in London over the past two decades and was successfully re-elected to the Greater London Authority alongside the Mayor, Ken Livingstone, in June 2004.

She is the Mayor's strategic planning advisor and has lead responsibility for the environment, including the setting up of a new Climate Change Agency for London. Nicky is also responsible for children and families.

Breakout Group Biographies

ROY COLVILE

Dr Roy Colvile is Senior Lecturer in Air Quality Management at Imperial College London, Department of Environmental Science & Technology. His work on air pollution in the past has covered every lengthscale from global climate via regional acidification to its current focus on local air quality. He is a leading investigator and was the original proposal main author of the DAPPLE project on Dispersion of Air Pollution and Penetration into the Local Environment, which focuses on determinants of personal exposure to traffic-related air pollution at microscale (1 m to 100 m) at the intersection of two building-lined streets. A member of the International Society for Exposure Analysis and Fellow of the Royal Meteorological Society, Roy's aim is to place the individual at the heart of urban environmental management, focusing on what individuals breathe and experience at the same time as considering how their activity determines the quality of the air they have to share with their neighbours in the increasingly crowded urban environment.

DAVID CROWHURST

David Crowhurst is Director of the Centre for Sustainable Construction at BRE. He leads a team of building professionals, environmental scientists, technical and administrative staff working on all aspects of sustainability in relation to the built environment: materials, buildings, communities and organisations. The particular focus for his own work has been on the development and use of indicators and facilitating sustainability management in organisations. He has worked with groups from Government, Local Authorities and the Private Sector — helping them to progress towards more sustainable construction.

JAN HEWLETT

Jan Hewlett is a member of the Greater London Authority's environment team, where she is particularly involved in work on biodiversity in relation to London's parks and green spaces. She chairs the House Sparrow Action Group of the London Biodiversity partnership. She is also currently president of the London Natural History Society and edited its recent bird atlas '*The breeding birds of the London Area*'. Jan can be contacted by e-mail as jan.hewlett@london.gov.uk.

RICHARD LAYARD

Richard Layard is Emeritus Professor of Economics at the London School of Economics, where he was until recently Director of the Centre for Economic Performance. He now heads the Centre's Programme on Wellbeing. His main work has been on unemployment and inequality. He was the architect of Labour's New Deal for the unemployed and is now a Labour member of the House of Lords.

JOHN MURLIS

John Murlis is a visiting professor in environmental policy at the UCL Department of Geography and a member of the Environment and Society research Unit (ESRU). He has over 20 years of experience of environmental policy in public bodies including the Overseas Development Administration, the Department of the Environment, HM Inspectorate of Pollution and the Environment Agency. His last public service role was as Chief Scientist and Director of Strategy in the Environment Agency.

During his time in the Environment Agency, Professor Murlis was responsible for a rapid expansion in the Agency's capacity to understand the broader public dimensions of its role, including on contentious licence applications, by increasing training, public debate and stakeholder dialogue.

Since rejoining UCL he has advised a number of high profile clients on environmental strategy, including energy companies, international industrial consortia, and learned societies.

Professor Murlis sits on a number of advisory boards for academic and public institutions and is the author or co-author of over 30 publications on science, environment and policy.

KEITH NOBLE

Keith Noble is City Birds Project Officer, based in the RSPB South East England Regional Office in Brighton. Before moving to Sussex, he was a keen member of the London Natural History Society, with a special interest in Rainham Marshes and the return of wildfowl and waders to the Inner Thames. He is now focusing again on birds and people in London. His subjects include house sparrows, herons and peregrines, and he is a member of their Species Action Plan teams in the London Biodiversity Partnership. Contact keith.noble@rspb.org.uk.

SIMON READ

Simon Read is Project Director at London Waste Action, with specific responsibilities for the London Recycling Fund and the associated London Recycling Project. London Waste Action aims to increase co-ordination between the producers of goods, retailers, manufacturers, the public and the statutory local authorities and all concerned with the management of waste and reusable materials and equipment, with the aim of achieving a more sustainable approach to its waste management.

TIM REEDER

Tim Reeder is a manager in the Thames Region of the Environment Agency. He has over twenty years experience in the environmental field, for much of that working to monitor and improve the quality of the Thames. He has been involved in climate change issues for over ten years and represents the Agency on the London Climate Change partnership. He is Project Scientist for the Thames 2100 project, which is looking at the future of the Thames Barrier and flood risk management in the Thames Estuary, and is managing the Agency's input to ESPACE an EC project

(European Spatial Planning Adapting to Climate Events). His e-mail address is tim.reeder@environment-agency.gov.uk.

CAROLYN STEPHENS

Carolyn Stephens is a Senior lecturer in Environmental Health and Policy and Co-Director of the Centre for Global Change and Health at the London School of Hygiene & Tropical Medicine, where she has worked since 1991. From 1999, she has also held an appointment with the Universidad Nacional de Tucuman (UNT), in Northwest Argentina, as a Profesora Titular en Salud Politica y Medio Ambiente. She is now also a visiting Professor in the Universidad Federal de Parana in Brazil. She has worked since 1981 internationally, as well as in the UK, in India, Liberia, Tanzania, Ghana, Brazil, Peru and Argentina. Her research focus is on environmental inequalities, environmental justice and health in developing countries, particularly in participatory projects with disadvantaged communities and children internationally. Most recently she has moved to work on participatory ways of using epidemiology supporting people to analyse their own health and environmental issues. Carolyn works with environmental justice and human rights lawyers and with farming communities, local urban groups and marginalised people in North and South. She has published widely and advises international agencies, governments, NGOs and community groups.

JENNY STOCKER

Jenny Stocker has been a Senior Consultant at Cambridge Environmental Research Consultants for four years. Her background is in the mathematical modelling of fluids, including high Reynolds number flows, and water waves; her current interests include air pollution as well as noise modelling.

Index†

---

†The words environment, sustainable development, and sustainability are so frequently used in the book that they are not indexed. Some of these topics are covered in the Appendix but are not indexed.